SO THIS IS FARMING

A Journal

by Dean Carlson

including an update on 1988, and
an addendum on sugar beets
and their processing

Illustrations by
P.R. Wagner

So This Is Farming

Published by Adventure Publications,
P.O. Box 269, Cambridge, MN 55008

Book & Cover Design by P.R. Wagner, Stanchfield, MN

Typesetting by Graphics Plus Midwest, Cambridge, MN

First Printing 1988

ISBN 0-934860-50-5 Softcover
ISBN 0-934860-51-3 Hardcover

ADVENTURE
PUBLICATIONS, INC.

To my future farmers: Melissa, Alan, Todd and Andrew

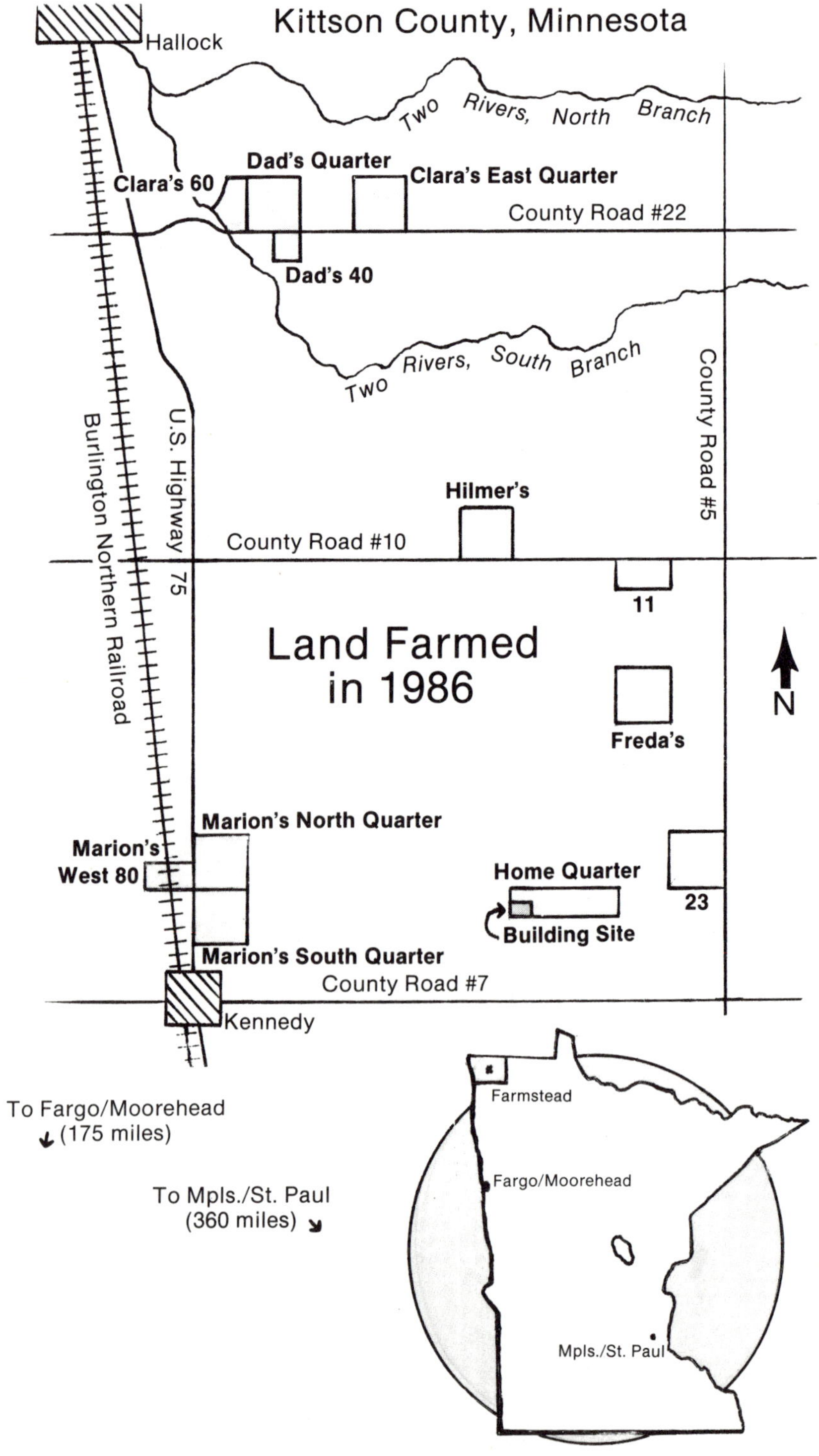
Kittson County, Minnesota
Hallock
Two Rivers, North Branch
Dad's Quarter
Clara's 60
Clara's East Quarter
County Road #22
Dad's 40
Two Rivers, South Branch
County Road #5
Burlington Northern Railroad
U.S. Highway 75
Hilmer's
County Road #10
11
Land Farmed in 1986
N
Freda's
Marion's North Quarter
Marion's West 80
Home Quarter
23
Building Site
Marion's South Quarter
County Road #7
Kennedy
To Fargo/Moorehead (175 miles)
To Mpls./St. Paul (360 miles)
Farmstead
Fargo/Moorehead
Mpls./St. Paul

PROLOGUE

I consider myself a typical Red River Valley farmer of 1500 acres of small grains and sugar beets. In 1986 I kept a journal of my daily activities. Handily tucked in the pickup's doorway cubicle, the diary secretly recorded my actions, descriptions and thoughts, as well as conversations with other people. (Changed names are indicated by asterisks.) Often contrasting with the news media's sometimes overly romantic and dramatic presentation of farmers, this book presents in accurate detail the occupation of plant husbandry.

One of my goals is to help readers understand the farmer's decision-making process. Conversations with neighbors portray the typical concerns and anxieties encountered in planting, raising and harvesting a crop. The unique problems, fatigue, and stress caused by the previous wet fall on the current season are shown to underscore the long-term effects of nature's influence. Frustrating visits to the local ASCS office, hiring beet harvest labor and watching the crop grow are all experiences in which the reader can share.

Beginning with spring's machinery preparation, the book shows the fleeting growing season of northwestern Minnesota. The long days of seeding and nurturing the young crop give way to a brief midsummer lull. Wheat combining and beet harvest complete the annual cycle.

As I began writing this book I wished that a more typical year could have been used. But what constitutes a "typical" year? I presumed it's one in which the crop is seeded early into a dry seedbed. With timely rains it grows and matures. Combining is accomplished without breakdown and with a minimal amount of artificial drying. Beets are dug under dry conditions without pulling trucks. Astonished at my foolishness, I realized that this is the exceptional, not the typical year. Every year brings its own peculiar circumstances and we remember each by, "That was the dry year;"

"That was the wet spring;" "That was the year of the four-inch rain during beet harvest;" "That was the year the combine caught fire;" "The army worms ate the barley that year;" or "The crop was terrific that year."

With time as the great teacher, we tend to develop the gift of keeping things in perspective. Things that worried me in 1986 don't bother me as much today. I can only hope the learning process will continue.

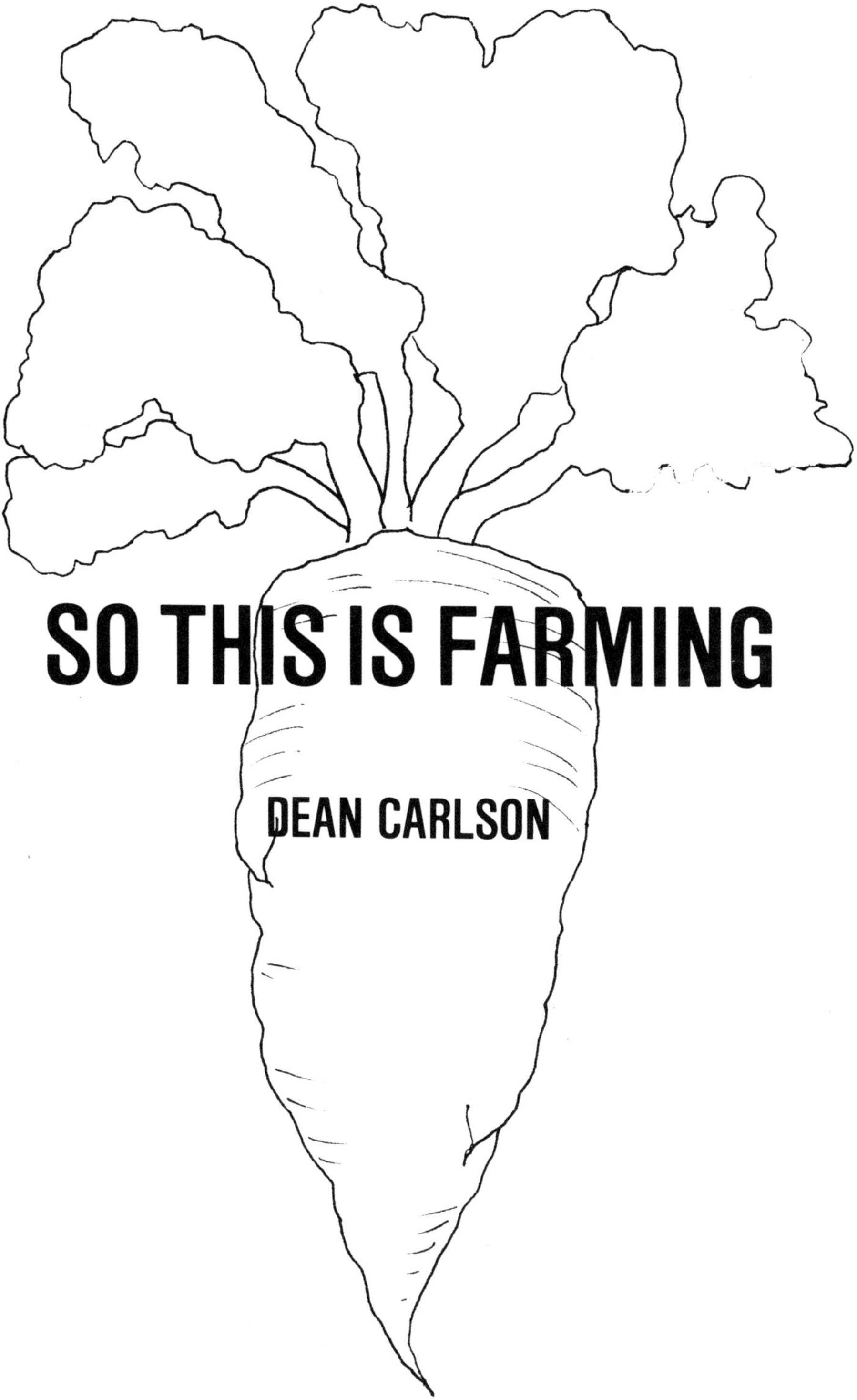

SO THIS IS FARMING

DEAN CARLSON

THURSDAY, MAY 1

At first I didn't hear the barely perceptible sound. Lying under a truck in the Quonset straining at a stubborn crankcase drain plug, I suddenly became aware of the arrival of the heralds of spring. I dropped my wrench and scrambled to the door to watch the Canada geese migrate. The spring air was crisp as I squinted into the morning sun. They were straight overhead and the whole countryside could hear their calling. I stood for some time watching them fly toward their summer homes in the Arctic. Another flock appeared in the west with another following, their formations constantly fragmenting and rebuilding. To the east a smaller flock came into view. Gracefully beating the air with their long wings and aided by a light tail wind, they too were making good time.

"They've got it made," I wistfully said to myself. Wishing I could go with them, I jealously watched as their V-formations disappeared on the northern horizon. I picked up my wrench and went back to work.

FRIDAY, MAY 2

The want ad read, "40 foot John Deere field cultivator in good condition." A name and phone number. "Pekin, North Dakota." Being a shopper and a perpetual searcher of bargains, I gave the guy a call. I was looking to trade off my old one or buy a newer one outright.

"Sure, it's in excellent shape."

"I'm driving quite a distance to look at it, so I would appreciate you telling me if there is anything major wrong with it. Has it been broken or welded upon? In what shape are the harrow sections? How are the shovels?"

"There's not a thing wrong with it. I'm going to no-till so I don't need a cultivator anymore. Oh, I guess the harrow sections on the one end are a little bent. We hit a dead furrow once, but nothing serious."

"And the shovels?"

"Good. Not a bit wore."

"The frame has never been bent or broken?"

"No."

"Well, it sounds like what I'm looking for and in my price range. If it's what you say it is, I'll take it. When would you be home?"

"How about tomorrow afternoon?"

"Fine, how do I get to your place?"

Having done this quite a bit, I've found most people are honest and polite enough to tell you of any serious flaws over the phone. There are, however, a small percentage of sellers who think nothing of wasting someone else's time and gas. The reasons for selling are varied. In this case, the farmer was discontinuing the practice of tilling the soil and would begin planting the crop directly into the previous year's stubble without touching the ground after the removal of the crop. Chemicals, instead of the cultivator, are more heavily relied on to control weeds with the no-till method.

Some of the other reasons for selling are: "I'm cutting back my operation;" "I got a bigger tractor and need a bigger implement;" "It's just sitting here cluttering the yard so decided to sell;" "I need to raise some money." I've yet to hear someone tell me that the reason for selling is that it's a piece of junk and the local implement dealer won't even take it in on trade. Unfortunately this was one of those times.

The harrow sections were bent, and those that weren't bent were broken. The shovels were completely shot. One wheel had broken off at some point and had been fixed by someone with pretty questionable welding abilities. After driving 140 miles to look at a cultivator that deserved to be melted down, I was disgusted. He either thought that I wouldn't notice the flaws or that after driving so far I wouldn't want to return without buying. Either way, he was wrong.

Sarcastically I asked, "Was anyone hurt in the accident?"

"Huh?" he replied.

I changed the subject, my point made. We chatted for a few minutes while I stretched, and parted with friendly goodbyes and good lucks. Chalking the trip up as just part of the game, I had found out where Pekin, North Dakota, is and unfortunately, quickly developed an unfavorable impression of its populace.

As it turned out, a neighbor traded for a new cultivator a week later and I bought his old one, driving only a mile to pick it up.

Russell Youngren, a farmer I worked for when I was in high school, made a deep impression on me and my farming career concerning machinery. Russell farmed a great deal of land economically; that is, he bought used equipment and plowed his profits into land, not into payments to implement companies. At the end of each year he didn't have to pay huge amounts of principal and interest for machinery purchased—I try to follow his example.

SATURDAY, MAY 3

The conversation was jovial at the local fertilizer dealer. Half a dozen farmers, dressed in dirty work clothes and coveralls, with mud on their boots, were sitting on the soft chairs sipping the strong black coffee and good-naturedly lamenting their problem at hand: the late wet spring. All were in their forties or fifties and had seen it all before. I drew half-a-cup of coffee and sat down to listen.

"Quite a change from last year," one commented.

"Guess we've been spoiled by early springs the last couple of years," answered another.

"Yeah, but you know, the ground isn't fit yet. I was walking out by my house and the land was cold. I believe if you planted anything it wouldn't grow yet anyway."

"Our machinery isn't ready yet either," smiled another.

Laughter from the crowd. The one commenting on the cold ground flushed as he sipped his coffee.

"This year, at least we'll be ready," added another.

"Some years we get the whole crop in and still don't have the machinery ready."

More laughter.

"I hear they're starting by Grand Forks."

"Beet ground?"

"All over from what I understand."

"We're the wettest spot in the whole valley."

All heads nod in affirmation. Eyes involuntarily turn toward the window.

"1950 we started seeding on the third week of June."

"How was the crop?"

"Not too bad, considering."

"It can happen."

The conversation became a little more serious, but was sprinkled with enough guarded optimism that upon leaving I felt good about my occupation and those who choose it as their living.

WEDNESDAY, MAY 7

Last November one of our 11,000 bushel storage bins had been filled with undesirably wet wheat. These bins, twenty-seven feet in diameter, were built with metal tubes beneath the concrete floor. The unloading tubes connect a hole (unloading sump) in the bin's center with an opening on the outside of the bin's foundation. Auger spirals are inserted into the tube when unloading and the grain will empty from the bin's middle to a waiting auger on the outside.

Aeration fans had cooled the grain for proper storage during the winter. As the spring's temperatures rose, I attempted to dry the grain by utilizing the fans. This method would have worked if I had left the fans on longer, but with no fresh air percolating upward through the bin, the wet wheat started to heat. This warmth created an ideal environment for molds, and a six-inch crust of spoiled grain developed. The problem wasn't discovered until today, when we started to unload the bin and chunks of spoiled wheat clogged the unloading sump after one-and-a-half loads. The only solution, regrettably, was to crawl inside and shovel the remaining encrusted grain out the manhole.

Armed with aluminum shovels and paper masks, our hired man, Robbie, and I climbed the exterior ladder to the eaves. Knowing we would be working up a sweat, we shed our jackets in the cool morning air. Rising from the interior, warm air steamed out the manhole.

"I don't think I'm going to like this," I told Robbie.

Moist air greeted us as we descended into the darkened cavern. Standing within a ten-foot crater of wheat, we surveyed the upcoming task. Chunks of spoiled wheat would have to be thrown out the overhead opening.

"It's only a few inches deep," commented Robbie through his mask.

Driving his shovel into the crust, he broke off a chunk of moldy grain and threw it upwards, toward the open hole. Most of it cleared the opening while some banged into the bin's tin roof. As a cloud of green dust settled upon us, I realized we were in for a dirty job. Without talking, we attacked our task. Slicing shovel-sized lumps of the moldy grain, we threw them out the hole. The warm, moist grain wet through my boots. Periodically one of us would miss our target, filling our working space with more of the dusty spores. Perspiration stuck my shirt to my back. I looked at Robbie through the haze and his once-white mask was black with dust. Sweat was pouring down the sides of his face. I must have looked the same. Wet with sweat and rotten grain, I threw my gloves out the hole.

After an hour I looked at Robbie and panted, "Good enough. Let's get out of here."

He didn't argue as he quickly grabbed the ladder and climbed out the opening. Sitting on the roof for several minutes, we threw our masks off and filled our lungs with fresh air. My wet shirt felt cold on my back as I sat head-in-arms.

Expectorating over the side, Robbie said, "Let's not do this again."

"Sounds good to me."

SATURDAY, MAY 10, 8:30 P.M.

The anvil-shaped thunderhead appeared in the southwest. I carefully studied the dark cloud with its trail of dark-looking rain falling to earth. The evening was calm, and the thunder could be heard clearly as the storm approached. Gradually, however, its southwest-to-northeast course became evident and showed that it would track to the west of us.

"Looks like that one is going to miss us," I said to myself. Someone in North Dakota received an unwanted rain. It couldn't have taken a very broad path and I was glad that we weren't in its way.

The sun came out in a brilliant flash of orange, sandwiched

between the black horizon and the dark, angry sky. In a moment it dipped beneath the horizon in a spectacular display of orange, yellow and purple. I considered myself fortunate. Later in the evening another cloud found us, and dumped .57 inch of rain before moving on. Looks like we'll have to wait a few more days.

I don't mind waiting, knowing that when the time comes we'll be ready. The tractors sit silently in the yard, fueled, serviced and coupled to implements; one to the cultivator, one to the drill and another to the beet planter. All are awaiting Mother Nature's signal to commence work for another spring. I find myself walking among them, checking and re-checking known weaknesses: a slightly worn bushing here, a chain that could possibly go another year there, items that aren't worth fixing but deserve to be watched during operation. It's a good feeling, knowing the equipment is ready. To be sure, there'll be breakdowns, but at least all the known repair work has been taken care of.

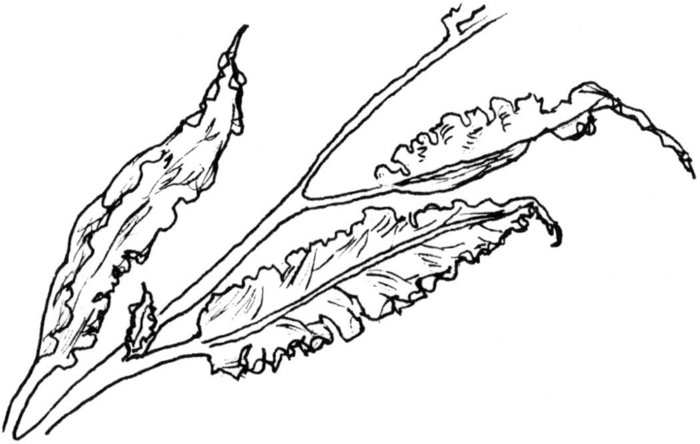

TUESDAY, MAY 13

The air was fresh and cool as I walked on Freda's quarter, 160 acres rented from a widow named Freda Stromgren. In most places the ground was soft. Skirting the dark gray wet areas, I stayed on the higher, drier land, examining for weeds and deciding when it could be cultivated. Thoroughly enjoying myself on this perfect spring day, I walked along trying to ignore the mosquitoes and kicking frequently into the ground. A pair of crows had built their nest a quarter of a mile south in a small grove of trees and the gentle southerly breeze was lofting their cawing in my direction. Constantly swatting the mosquitoes, I quickened my pace. My main concern was the unusually large population of curled dock, also called sour dock or Indian tobacco. If left untouched it can grow anywhere from one to four feet tall and will choke any other plant within two feet. Its brown, triangular seeds resemble dried tobacco. Occasionally I stopped and pulled one out of the ground. They were already developing sizeable taproots.

With most weeds, the time to kill them is when they are young. A couple of deep cultivations with a chisel plow would do the trick, but not on land to be planted in sugar beets. A chisel plow is not a very good tool to use in the spring of the year for seedbed preparation, as it leaves the ground too lumpy and will dry the ground too much for good seed germination. It would, in this instance, dig the curled dock out, but the results of leaving a terrible seedbed would be entirely unacceptable.

The logical alternative is to spray by air. But how many acres? I'm calculating costs as I estimate how many acres would have to be sprayed. At $6.75 per acre should I spray the whole quarter, or only where the weeds are the thickest? The field is wet. How many days of good weather will it take to dry? What's the weather forecast? The beets should be in the ground by this time. How much yield loss can be expected from the curled dock? What would be the cost of hiring labor to weed out the dock after planting? I am contemplating these questions as I walk. If there were only one unknown in the formula, the decision-making process would be easy. In this situation an educated guess will have to do. The pickup waits for me as I plod in its direction. Jumping the wet ditch with a splash, I head for the local dealer to get the chemicals.

WEDNESDAY, MAY 14

Since the machinery is ready to go, I spend a lot of time in field inspections. Today was no exception. Last year's beet ground will be the first to be cultivated this spring. Sugar beets absorb more moisture than wheat or barley, so the soil is dry for the next year's crop. With this in mind I drove the three miles north of the farmstead to Hilmer's quarter, rented from Hilmer Bengston's widow. Most of this particular quarter is a light gray color, with the exception of a few wet holes. Stopping the pickup occasionally, I walk a few steps, driving my heel into the earth. The aroma of decaying sugar beets and tops fills the air.

"No reason why this shouldn't go," I thought.

Urea fertilizer will be incorporated this spring. Normally all the land would have been fertilized with anhydrous ammonia the previous fall, but with the terribly wet conditions in 1985 not much was applied. Anhydrous ammonia can be applied in the spring, and a few farmers follow this practice. However, on our clay soils when anhydrous ammonia is applied, the knife applicator digs into the ground and has a tendency to dry the soil. Also, the nurse tank being towed behind the applicator is extremely heavy and will leave a set of tracks on the field. Anhydrous is 82% nitrogen while urea is only 46%. Urea is a pelleted form of nitrogen which can be spread on a field and cultivated prior to seeding. It is more expensive, but most farmers in the spring of 1986 chose it because they didn't want to dry the soil and didn't want the tracks. Ideally the anhydrous ammonia should be applied in the fall since the freezing temperatures will keep the fertilizer intact for next year's crop.

I drive on, zigzagging the field. Drawing a map of the quarter, I note the wild oats peeking through on the coulee banks and the kochia weed appearing on the west side. Having satisfied myself that the whole quarter could be worked, with the exception of two small holes, I leave for home to call Cenex.

A major job this winter has been to study the farm program with its many options. About the time I get it figured out it changes, forcing me to recalculate. It's necessary to make contingency plans in case of a wet spring or if a particular piece of land can't be seeded. Farm maps are drawn, and fields are measured and staked where they are to be split, part into wheat, part into summer fallow, or set-aside or ACR (Acreage Conservation Reserve) or whatever new term the government can come up with to call it. Each farm is assigned a wheat base, depending on its individual past history of how many acres of wheat have been raised, and a certain percentage of the wheat base must be taken out of production. This year it's twenty-five percent. That is, if you had a 1000 acre wheat base, you'd have to summer fallow (ACR) 250 acres. Failure to do so results in your losing all government payments and subsidies. The government can control the number of seeded acres of wheat, and to a lesser extent, the country's production. Appeals are made to the ASCS office to increase one's wheat base or proven yield. Figures are added and re-added to determine that one is not over-seeding while maintaining enough fallow.

Also during the winter, seed is cleaned or bought. Machinery and tractors are serviced and greased. All these preparatory activities come to mind as I watch the cultivator wings slowly unfold.

A forty-acre tract of lighter ground on Section 23 may be fit today. Situated on a coulee bank with good drainage, it traditionally is the first to get worked. Farmers call this type of land "lighter," meaning it is sandier soil with a higher percentage of sand to clay. The term "light is a bit deceiving in that a cubic foot of this type of soil weighs more than a cubic foot of clay. Maybe we call it light because it is easier to plow. Farmers also call the land more forgiving, in that you can work it wetter and not get the wheel track problem associated with a soil composed of more clay.

One of the toughest decisions in farming is when to start cultivating in the spring. If worked too wet, our clay soils develop impermeable tractor wheel tracks. Not being able to adequately penetrate these tracks, the grain drills shallowly plant the seed into dry dirt. A timely rain may dissolve the tracks and wet the seed for germination, but if it doesn't rain black wheel tracks of unsprouted grain are the result. If one is too timid about creating tracks and waits too long to cultivate, rain delays could hinder seeding's progress, resulting in later planting dates and reduced yields.

Different farmers have varied means for deciding when to start: "I don't start if I can roll a mud ball behind the cultivator;" "I like to see the first spears of wild oats start to appear;" "I don't want it to be sticky;" "I wait until the tractor doesn't compact the ground;" "I wait until the neighbors start." Over the years I've been hurt more by waiting than by working the ground too wet, so I decided to work.

When it's the middle of May one must take a few chances.

Aligning the radiator cap with a distant power pole, I drive down the field, taking a forty-foot swath. The first two rounds go slowly, as I'm constantly getting in and out of the tractor, walking behind the cultivator in the newly-worked ground. If the calendar read April 14 instead of May 14, I would wait. I'm feeling that it's too wet, but keep going. Looking across the landscape, I see one of the neighbors has also come out. That's better. If I'm doing the wrong thing then we both are. No doubt he's thinking the same thing.

Barley will be planted here and perhaps sugar beets in 1987. It'd be nice to get this barley seeded and harvested early in order to get the ground worked properly this fall for a good seedbed for next year's beets.

This is the most relaxing part of farming, to steer a tractor up and down the field and daydream the afternoon away, sing with the radio or plan tomorrow's, next week's or next year's events without distraction.

About 6:00 a shower comes from the northwest and puts a halt to the first day's cultivation.

THURSDAY, MAY 15

Mud was sticking to the pickup tires as I drove north to Hilmer's quarter. I kept listening, hoping that the sound wouldn't get any worse. Relieved to find that only about a tenth of an inch of rain had fallen during the night where the fertilizer spreading would be done, I drove home for breakfast. The rising sun promised a beautiful day in the making. To the west the refracted light produced the often-seen mirage of the Red River fifteen miles away. I could distinguish building sites along the river that ordinarily were not visible. A terrific sight. I stood in the yard for some time admiring the morning. The wind was calm. In the distance a tractor could be heard, interrupted occasionally by meadowlarks.

Robbie finished cultivating on Section 23 and moved to Hilmer's to incorporate the freshly spread urea. The wind rose from the southwest to create a beautiful drying day.

About 7:00 I pulled the drills to 23 and started seeding Robust barley. As I was seeding, I again wished I'd waited another day or two. This will be a good test of how wet we can work this field and still get a stand. Clods of mud had been left behind the cultivator and were now drying into hardened chunks of soil.

FRIDAY, MAY 16

The Roundup jugs slid forward as he braked his pickup. "Got some quack, Byron?" I asked.

"Yeah, about eighty acres," he replied with a smile. "Are you going?"

"We are," I answered. "But I'm not so sure we should be. There comes a time when you just say to heck with it and start."

"We aren't going to start until Monday. But with the middle of May, maybe we should go, too. I don't know." He rested his hands on the steering wheel and gazed around the countryside.

We talked for several minutes, two farmers sitting in their pickups on a country road in the spring, leaning out the windows discussing seed varieties, weather, farm programs and tractors.

The pickup and I must have driven a hundred miles today just inspecting fields. Everywhere I check it's the same condition: too wet. A few good drying days with a good southwest wind could change things quite a bit.

Robbie finished the thirty acres on 23, vacuumed all the remaining barley seed from the drills, moved to Hilmer's (a distance of seven miles), and started to seed wheat. This is the first year that he's running the drills extensively. I get a little nervous having someone else doing the seeding. I keep wondering if the drill is in the ground properly or if the seed and fertilizer are being metered correctly. However, the time has come when I have to trust someone else because I'm busy planting and caring for the sugar beets. Robbie comprehends quickly, and after one or two fills became quite annoyed with my constant lecturing and surveillance. I was trying to explain the series of steps I go through to fill the drills, which saves a few seconds here and there that ultimately add up to minutes.

Being a man of few words, Robbie listened quietly each time but finally burst out, "I think I can do it faster this way!"

I watched as he developed his own routine, and sure enough, it was faster. I left him alone after that. Robert Peterson has been excellent hired help the last five years. He is just twenty-three years old, but has matured into the type of man who can do anything around the farm. I sometimes think that the place would fall apart without him, and don't mind telling him so. It's a relief to know that he can seed and spray just as well as I can and that while I'm band spraying or cultivating beets, he's doing a good job with the grain.

I was attempting to burn some of last year's wheat stubble that didn't get combined. Watching the orange tongues of fire lap at the crisp straw, I realized once more what a loss the previous wet fall was. The gusty wind would race the flames forward at a fast walk or slow them to the point of dying until a fresh gust would carry them along again. Behind was left a black, charred ground mottled with the heads too damp to burn. I was amazed at how many heads there were. But 1985 was the best wheat crop I'd ever raised, so it shouldn't come as any surprise. I walked in the cinders and kicked at the burnt straw, exposing more heads. The northwest wind felt cold as it carried the cinders aloft.

"Dirty shame this couldn't have been combined," I said to myself. Then I remembered an old neighbor, Clarence Hagen, who once took me aside and gave me some fatherly advice when I started farming in 1973:

"That's the way farming is," he said. "You have to take the good with the bad."

I always wanted to farm but never had the resources to start, since so much capital and land are required. My dad owned 200 acres of land southeast of Hallock, but since it was too small a piece to make a living on, he rented it to a neighbor.

After I graduated from NDSU, a good friend of mine, Marvin Bengtson, offered to rent me some land and machinery. Since he had no one in his family to help him, I would trade my labor in return.

In the spring of 1973, I tried to borrow operating money to get started, but discovered a snag—no creditor would touch me without collateral. Marvin graciously offered to loan me $5000 to buy seed, fertilizer and chemical. I still have a copy of that paid note as a reminder of the time when $5000 would put the crop in. I owe my farming career to Marvin's generosity.

I started that year with Dad's quarter and Dad's 40, the quarter on Section 23, and the eighty acres on section 11, and have acquired new acreage to own or rent ever since then. I've had ups and downs in these thirteen years. In 1980 I lost a crop to drought and my bankers weren't smiling too much. But I believe that losing that crop helped me put things in perspective and helped me grow as a person.

"How right you are, Clarence," I said out loud as if he were standing next to me. "Take the good with the bad."

The sun was going over the horizon in the west as I crawled into the pickup and headed for home.

SATURDAY, MAY 17

One of those intangible rewards farming has over other occupations is the privilege of smelling freshly turned soil in the spring of the year. Cultivating on the home quarter and getting out of the cab frequently to adjust the cultivator setting, I would pick up a lump of dirt, hold it to my nose, and with a series of shallow breaths enjoy the smell of spring.

Years ago in a soil microbiology class at North Dakota State University we were trying to isolate a certain bacteria. Upon coming to class, we all opened our petri dishes to find the whole room filled with the smell of a freshly-cultivated field. One of the students who obviously had a farm background commented on it, which immediately brought a smile to the professor's face.

"Exactly," he answered. You've just isolated the bacteria responsible for that smell."

The morning was warm and sunny. I thoroughly enjoyed looking back and seeing the light gray soil turning black at twenty-five acres an hour and quickly finished the small piece. I spent most of the afternoon checking fields. The two fields that concern me most are Freda's and Dad's 40, which will be planted to sugar beets.

I walk Dad's 40, stopping occasionally to dig in the rich, black soil. The land has dried considerably the last few days. We'll bring the multiweeder here on Monday, and hope to get some beets in the ground by Tuesday.

Bumping along in the pickup in a rough field near Hallock, I stumbled upon another reminder of last year's wet fall: two very deep tracks cut through a coulee which meanders through this quarter. It was here that a tandem truck loaded with wheat got buried. After breaking two chains trying to pull it out, we left the truck for the night. The next morning we brought an auger to unload half the wheat so the truck could be towed from the mud. That wet fall will continue to plague us all spring, as a poor job of seedbed preparation is being done.

I walked in a stubble field observing the ruts from the previous fall. Winter's freezing and thawing had softened the ridges somewhat, but still a well-defined, deep track could be seen where a truck had been stuck or where a combine had to be pulled backwards out of a hole.

From the pickup window I could see deep ruts twenty feet apart running the whole half-mile length of Harold's field.

I saw Harold inspecting his badly-tracked field and stopped to talk.

"Think this'll make a seedbed?" Harold asked facetiously, walking up the ditch to where I was standing.

Trying to maintain a positive front, I cautiously replied, "It'll take a couple of cultivations."

"You know what'll happen, don't you? We'll work the shit out of this," he said, waving his arm in the direction of the field, "dry it out, seed it and not get a rain."

His comments reflected the feelings of most farmers forced to seed into such a poor seedbed.

"There could be some erratic stands," I agreed.

"Well, there won't be a thing," he answered. "Doggone it, we've gotta fill those tracks in somehow. Oh, I tell you, Dean, this farming is strictly a no-win game. Every year something happens. I had a beautiful crop last year, but after I paid for the custom combiners with rice tires and the repair bill, I'd have been just as well off with an average crop. Gotta cigarette?"

I shook my head.

"I've been farming twenty years now," he continued, "and I bet I can count on one hand the number of years that I made money. Chemical, fertilizer, rent just keep going up. New machinery's completely out of line."

I found myself acting the role of the good listener as he rambled on about the ills and pitfalls of farming. Occasionally I would attempt to stifle his oratory by submitting a positive point or two, but his mind was made up. I found myself being outflanked time and again by his pessimism. After several long minutes of this narrative he smiled at me and said, "Well, have a nice day."

"Have a nice day? You just ruined mine!"

We were both laughing as we drove in separate directions. A psychologist would have charged him a hundred dollars to get all that off his chest.

MONDAY, MAY 19

Warm, southerly winds predict an excellent drying day. Just the kind of weather we've been waiting for. It's going to be a spring of risk-taking and experimentation. I've never before had to plant a crop into such a poor seedbed. Weeds are especially prolific this spring, probably due to the lack of cultivation last fall and summer. Aside from the usual pigweed, smartweed, buckwheat, kochia, lamb's-quarters, wild oats and curled dock, there are large populations of new varieties I've never seen before. The county agent and local chemical dealers have their copies of *Weeds of the North Central States* out, trying to identify them.

These varied weeds were giving me problems this morning. Robbie had made one round with the multiweeder on Dad's 40 in anticipation of planting beets. Dad's 40 is a piece of ground touching Dad's home quarter. It is rented from my dad. I walked the freshly-worked ground, kicking and digging with my foot. Everywhere were weeds that had slipped through the multiweeder's tines. Many times some dirt had been thrown on top of the weeds and they had not been torn out. Usually a multiweeder is an excellent seedbed preparation tool in that it lightly cultivates the ground while attached harrow sections smooth the soil for planting. Much to my dislike, we'll have to cultivate this piece and run the risk of digging it too deep and creating a very lumpy seedbed for beets. I can't see planting into a field dirty with weeds.

Seedbeds for wheat and sugar beets differ somewhat. Wheat can be planted into a rougher, lumpier seedbed. The wheat kernel is larger and is planted deeper into the soil where it reaches moisture. Both wheat and sugar beets require a somewhat firm seedbed, but beets need a firmer one. There are times when we drive our empty drills over a field before seeding beets in an attempt to firm the soil. Having a smaller seed, the beets will not emerge if planted too deep; therefore, they must be planted shallow and into moisture. If the field has been firmed, good soil-to-seed contact will be made and the seed should sprout. Lumps must be eliminated. If the shallowly-planted beet seed is under a dry dirt lump, it will not germinate.

Harvest States Fertilizer Company has spread quite a bit of urea for us today. We'll be kept quite busy cultivating it into the soil before warm temperatures cause it to volatilize (become vaporous.) Robbie finished cultivating Dad's 40 and started on land that will be seeded to wheat.

In a futile attempt to break the lumps, I harrowed the beet ground. Around 5:00 I started to plant sugar beets into a rough seedbed. Lumps of wet, soft dirt had dried into stone-hard clods which kept lifting the planter units.

Looking back from the tractor seat, I could see, to my chagrin, the tiny, bright-orange beet seeds lying in an opened furrow, completely uncovered. The constant stopping to adjust the setting made it a frustrating first-day's planting. The field varies so much that what may be a proper setting for one end is too shallow on the other. I'd like to plant three-quarters to one-inch deep, but I'm finding that it's almost impossible. Planted that shallowly on this particular field, all the seed would be in dry dirt, needing rain to germinate.

Across the road Bob Daigre was cultivating for me in the 850 Versatile. Bob was hired for the spring rush, and is excellent with machinery. Having promised his labor to a custom combiner, he will be with us for a very limited time. I'm anxious to get as much work out of him as possible before he has to leave. Gifted in the verbal arts, he's constantly jabbering and offering his advice over the CB radio.

"Those beets'll be coming up before you get done planting," he laughed over the air.

I didn't answer. I was totally frustrated. Two tools of the trade for planting beets are a tape measure and pliers. Using what is commonly called the beet grower's stance, knees and elbows on the ground straddling a row, rear end in the air and nose six inches from the ground, I probed with the care of an archeologist with the pliers' handle. Carefully flitting the dry dirt aside, I searched for the beet seed. Suddenly, like a bright-orange jewel on a black pillow, the beet seed appeared. Meticulously digging around the seed in search of moisture, I would find in most cases that I had managed to press the seed into the wet earth. Other times dry, crumbly dirt surrounded the seed, which meant a rain would be needed to bring life.

The tape measure is used to measure seed depth carefully because depth is important. The planter's depth bands are round metal wheels attached to each planter unit. Their purpose is to keep each unit from penetrating too deeply into the soil. Each one of the twelve planter units has two depth bands on either side of the double disc openers. These discs slice a furrow into the soil into which the seed is dropped. If one of the depth bands is removed it will allow the discs to penetrate somewhat deeper. If both are removed the planter will plant the seed too deep. Having removed the depth bands to achieve greater penetration, I found that although the seeds were in moisture they were dangerously deep, perhaps too deep to reach the surface. I weighed the alternatives and decided to take a chance on planting too deep rather than placing the seed closer to the surface. Finally after four-and-a-half hours of planter adjustment and dozens of trips in and out of the tractor, I managed to plant fourteen acres of sugar beets. Not expecting a very good germination percentage, I spaced the seed at a thick three inches, very close.

That translates into 400 seeds for every 100 feet of row, so if they all grew that would mean 400 plants, much too thick. The beets would be the size of carrots. The correct number of plants at harvest is 120 to 140 plants in 100 feet of row for optimum size and sugar content. In 1986 our average number of plants before thinning was 200 plants in 100 feet of row. Where did the other 200 seeds go? A few didn't germinate. Cut worms probably took a few. The wind took some plants and the dry, lumpy seedbed, preventing proper germination, probably took the highest toll .

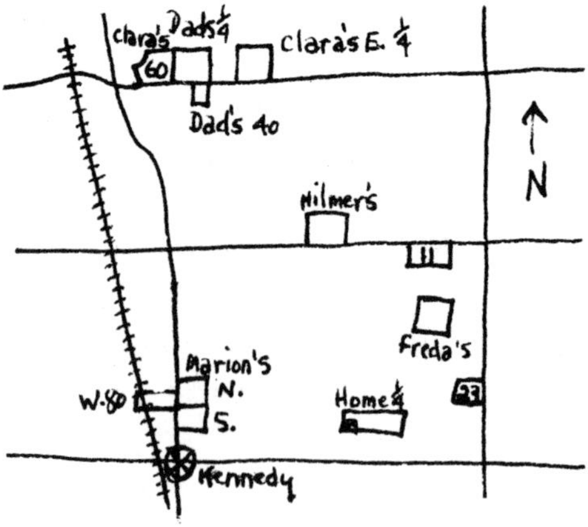

TUESDAY, MAY 20

Another beautiful sunny day. I finally finished planting beets on Dad's 40 and moved the planter to Freda's quarter. Robbie is doing a good job of seeding wheat in spite of, or more likely because of, my not badgering him. Yesterday he seeded a chunk on the Home quarter, moved the drills six miles to Hilmer's, seeded eighty acres and moved another ten miles to Clara's 60, rented acreage named after Clara Blide. Our land is spread out quite a bit, which can be a nuisance when transporting drills, but can be quite advantageous when a hailstorm tracks through the county. The odds of its hitting all the land are smaller.

Most machinery can be moved quickly from field to field. Bob finished cultivating Dad's quarter, left the cultivator at Hallock and brought the multiweeder the ten miles to Freda's quarter, where beets will be planted. I rode with him on the first couple of rounds, getting out frequently to inspect the land. We will have to go around five potholes, a nuisance for planting. We'll have to come back and seed them later, tying in the rows. Trying to keep ahead of Robbie with the drills, and me with the beet planter, Bob finished multiweeding half of Freda's, moved back to Clara's east quarter and cultivated the rest of the day.

Robbie is making good time with the drills.

"We're getting most of it in moisture," remarked Robbie as he dug in the ground.

The four-wheel-drive Versatile tractor idled while the drills sat. The two of us walked in the freshly-seeded wheat field. Stopping every few steps to dig, we could brush away the dry cinders and find wheat seed pressed into the wet ground two inches below the surface.

"The wheel tracks are dry though," he added.

We dug lightly in the tracks left by the seed tractor. Just beneath the surface lay the seed in dry dirt.

"I'm afraid the cultivator tracks won't come either," I lamented.

Sure enough, the 850 had also compacted the ground, creating tracks at a forty-five degree angle to the drills. The drill discs just couldn't penetrate the hard pan created by the heavy tractor wheels.

"I don't know what else to do," I said, putting my hands on my hips and gazing at the field. "We've got all the press on and heavy duty springs in the tracks."

"Just have to seed and hope for a rain," commented Robbie. "Too bad we couldn't have worked the ground a little more last fall."

I nodded. The land was only chisel plowed once, leaving it rough and hard for this spring.

"You wait. We all want dry weather now, but a couple of weeks from now there'll be a cry for rain," I prophesied. "Guess all we can do is keep on seeding and hope for the best."

Robbie climbed into the cab, opened the throttle and proceeded down the field. I watched until he was a quarter of a mile away and the squeak of the drills and the roar of the Cummins diesel could no longer be heard. A plume of black exhaust smoke drifted to the north.

WEDNESDAY, MAY 21

We're having a string of beautiful days. I hope it holds. Everywhere are red, green and white tractors grooming diagonal paths across unworked fields. It's an exciting time of year. Robbie finished seeding wheat on Dad's home quarter to finish the day.

The John Deere purred at three miles per hour as it drew twelve lines at a time up and down Freda's quarter. The packer wheels pressed the beet seed into the warm, mellow ground. Planting conditions are ideal. It's a good feeling knowing every seed planted is getting into moisture.

It was not to last long, however, as the farther west I planted, the heavier and lumpier the land became. Once again I started to experiment and wasted much time crawling on the ground looking for the seed. Once again I adjusted and re-adjusted the planter. All too often the seed would be sitting right next to or right underneath a large clump of dirt. High and dry, it didn't have a ghost of a chance to germinate. I was spending more time adjusting and digging than planting. Nothing seemed to help.

If I knew we were to get a rain shortly I'd keep on planting. The weather forecast is for mostly sunny with a chance of showers for the weekend. The thought of seeding into dry dirt doesn't appeal to me. What to do? If a person likes to make decisions, he'll love to raise sugar beets. The sun was setting in a haze of orange as I quit and went home.

THURSDAY, MAY 22

"I can only work till noon today," Bob announced, walking into the Quonset.

"Why's that?" I asked.

"The guy called from Grafton last night and said they're leaving for Texas tomorrow morning."

It was true that I had hired Bob with the knowledge that when he had to leave with the harvest brigade, he could go.

"Do you suppose I could talk you into working the whole day?" I asked.

"Sorry, but I've got a lot of things to do. Put my pickup in moth balls for the summer, do some banking and some errands."

“Gee, we’re really in a bind, with the late spring and all.” I was talking so sweet that I couldn’t believe myself. “If you could get one more day on the 850, we could be caught up with the cultivating until I get the beets in.”

“I know, but I gotta go.”

“Could you at least finish the quarter you’re on?”

“I’ll do what I can until noon.”

“If I give you an extra fifty cents an hour, could you finish the quarter?”

“Well okay, but call my girlfriend and tell her I’ll be a little late.”

I had talked the universal language—the common denominator which all of the human species understands—money.

Every day Robbie is making good strides in seeding wheat, while I’m stuck at a snail’s pace planting beets. Today would be no different. I had heard of some growers mounting a set of V-shaped plows in front of their planter units to push the clods aside. With the dry upper layer of soil shoved aside, the planter could send the seed into a moist seedbed. I went searching for a set. Inquiring at John Deere if a set might be purchased, I was informed that the plows are not a Deere product but are homemade. I decided to scour the countryside for a set to rent. Coffee cup in hand, I drove the pickup down back roads looking for someone planting beets. Luckily, the first grower I ran into was Charles Swanson. Walking in his newly-worked field to meet the planter, I was impressed by the firm, pulverized seedbed.

“The seedbed looks terrific. What did you do to it?” I asked, picking up what I thought to be the biggest dirt lump I could find. Its size was no bigger than my thumbnail.

“Oh, I rented a seedbedder. It didn’t do too bad a job.”

We conversed about farming briefly while I inspected the plows on his planter. They were a different type from what would fit on mine. I was hoping to get an idea of how they were made so I could spend the rest of the day, if necessary, fabricating my own. Our conversation couldn’t have lasted more than three minutes. All the while Charlie was nervously jumping from planter unit to planter unit, pulling off the lids, inspecting and probing with a stick. Although very cordial and polite while he was talking, he nevertheless displayed a behavior not uncommon to farmers in the spring: eyes darting back and forth while looking up and down the field, fragmented speech, hands that can't seem to find a comfortable place to rest, feet that pace twice to the left, pivot and stride to the right. Although I appreciated his congeniality, I felt uncomfortable taking his time.

Five miles down the road I found Ronnie Anderson, paint brush in hand, sitting by his shop door painting the discs on his beet planter. Having finished yesterday, Anderson was already giving his planter its annual rust-inhibiting coat of preservative. Not one blade of grass was left uncut in his spotless yard. Every tractor and piece of machinery was sitting in its proper place. The shop was swept, with tools and supplies in their labeled cubicles. On his John Deere flex planter was mounted a set of plows. Explaining my dilemma, I asked if I could rent his units for my remaining eighty acres to be planted. He refused, but stated that I could borrow them instead. The deal was too good to pass.

Driving around the country on a beautiful spring morning, seeing everyone else planting, and realizing your tractor is standing can be an unnerving experience. I keep telling myself that if a proper job is to be done, a little more time will have to be taken. I hope the rain holds off a few more days. Mounting the twelve plows on my twelve row planter and experimenting, I found that they did indeed help, but there were still too many large lumps hindering the depth control.

After a couple of rounds I quit and went searching for a seedbedder to rent. Urbaniak Implement at Kennedy had one for three dollars per acre. I spent most of the afternoon picking it up and working the remaining eighty acres on Freda's.

One of those specialty pieces of equipment developed for the sugar beet industry, a seedbedder contains several rows of light duty cultivator shanks designed to lightly till the soil. An attached series of rolling drums compact the ground, creating the proper seedbed for beets. It did an excellent job. The Danish tines worked the soil while the rolling baskets crumbled and packed the ground beautifully. There was still enough daylight to plant a couple of rounds.

Today two items were added to my list of potential purchases: a set of plows for the planter and a seedbedder.

It's been a very tough few days for my wife, Carol. One by one the four kids have all come down with the chicken pox. Even Andrew, who is only four-and-a-half months old, has had them. Dealing with four irritable, sick kids is a chore. Carol is taking the disaster in stride with a terrific amount of patience. We took pictures of them all with their red pox over their faces and bodies. Alan accidentally scratched a pox on his nose, and I'm afraid he will have a permanent mark.

FRIDAY, MAY 23

The weather forecast was for a chance of showers this evening and again tomorrow. The sun wasn't over the eastern horizon when the John Deere was making straight lines down the field with the planter, twelve lines at a time, seventy-two lines per hour. The plows were working beautifully. Pushing to the side the dry dirt lumps, the planter's double disc openers came behind and put the tiny seed into moist, warm ground. Clear sky gradually gave way to light clouds appearing from the west. As the afternoon waned, these were followed by dark low stratus.

Finally about 7:00 p.m., I reached the other side of the quarter. Idling the tractor for several minutes, I sat and observed my handiwork, relieved to have the beets in the ground. Later, I drove the pickup along the full half-mile headland, admiring the straight, or at least mostly straight rows. Confidence welled in me: regardless of what nature would decide to do, I had done all I could.

Robbie finished seeding Clara's quarter and drove the thirteen miles down highway 75 to Marion's. Marion Rath rents three parcels of land to me; Marion's north quarter, Marion's south quarter and Marion's west 80.

Scott Ehrenstrom came after work to do a little moonlighting on the cultivator.

It was dusk when I met Robbie at Marion's. Something looked peculiar about the drills. Upon inspection I found two press wheels had somehow fallen off en route from Clara's. After a search, one was found in the ditch along the highway. The other was lost. I sent Robbie home, and called the local parts man at his home. He was very helpful, and I was grateful for his opening after hours. The drill was fixed and ready for seeding the next morning.

SATURDAY, MAY 24

While everyone slept, nature was doing her part in raising a crop. At first a few small, sporadic drops of rain fell on the dry topsoil, creating mini-explosions of dust upon impact. Slowly the drops began covering the ground until the surface was completly veneered. As more rain fell, the pores filled, saturating the thin topmost mantle. Gradually its downward percolation infiltrated new air caverns. The top two-inch profile, dried from cultivation, quickly absorbed the life-giving moisture. The moisture-hungry clay particles quickly absorbed and swelled, slowing penetration. The small trench in which the beet seed lay was the recipient of a higher percentage of the moisture. Seeping down the sides of this micro-valley, water sank deeply into the base of the incision, engulfing the newly planted seed. In a few hours the dry, hard epidermal seed covering would be softened and germination would begin.

The rain gauge read a shy .4 inch. I called Robbie to inform him to take the day off, and set out to inspect fields. Stepping into the beet field, I was pleasantly surprised. To my amazement most of the beet seed got wet and "moisture met" in the planted furrow. I couldn't believe my good fortune. It looks as though we're going to get a stand.

The grain did not fare so well. Because it was seeded deeper and on old ground, it didn't receive the little bit of moisture, which soaked only slightly into the ground before being absorbed by the thirsty clay. In the wheel tracks the shallow seed had received water but a deadly layer of dry soil lay underneath the wheat. Unless the rain continues these seeds will germinate and die, slashing the yield.

The clouds blew away as the wind switched to the northwest. I was reminded of the old maxim: "If you ever want the sun to come out, send the hired man home." I could have used Robbie to seed about 5:00 p.m. The weather forecast is for sunny skies and good drying conditions. If the weather remains fair, seeding will go better on Monday.

After supper I cultivated until dark on Marion's west 80, again going around several holes. Ordinarily I would have quit, but this late in the spring a little bit of mudding will be necessary.

SUNDAY, MAY 25

Sundays are a time for rest, worship, getting re-acquainted with my children, golfing, lying in the sun on a blanket with a good book, swimming, trips in the car, naps and driving slowly around the section in the pickup, observing the crop.

Accustomed to getting up much earlier, I woke before the 7:30 alarm. Church is at 9:30, so I had plenty of time to bask in the warm morning sun. One of the most relaxing activities in this world has got to be sitting on the patio, coffee in hand, listening to the meadowlarks on a Sunday morning. I'm extremely possessive of my Sundays and wish to leave them open for family activities.

While the kids were taking naps, I did afford myself a trip to the beet field. Crawling on hands and knees, digging for seeds, I was amazed at the sprouts. Germinating under seemingly impossible conditions, some already have three-quarter inch sprouts. Some seed is barely into moisture, but evidently it's damp enough to crack its hard coating and germinate. As each tiny shoot was found protruding from the seed, I found myself repeating, "Can you believe it?" "Well, I'll be darned;" "What a stand!"

On the way home I stopped to walk in the barley field. Its dark green color could be seen a quarter-mile away. Having been in the ground ten days, the young plants are in the one-leaf stage. The fat, healthy-looking leaves emerging in rows are a beautiful sight. I stood admiring the scene for several minutes, receiving a farmer's reward for his work.

MONDAY, MAY 26 MEMORIAL DAY

"Look at those lucky devils," a local farmer said, motioning with his head toward a camper pulling a boat down Highway 75. "Looks like all the city people have the weekend off."

Standing in line to get fertilizer at Harvest States, I could see what he meant. The highway was inundated with recreational vehicles and cars pulling campers and boats.

Leaning against his truck while the liquid fertilizer pump ticked off the gallons being loaded, he continued, "Wish we could spread our work load out over the whole year and get more time off now."

"Guess you're right," I agreed. "Too bad we can't enjoy the summer months more."

We stood for several minutes watching in silence. The hum of the pump and the compressing of truck springs were the only sounds. A farmer from west of Kennedy, he seemed lost in thought, but as the fertilizer approached the 1600-gallon mark, he climbed into the box, and with new resolve announced, "As soon as the crop's in, I'm gonna do a little fishing myself."

We had a good day today. The term "good", of course, has different meanings for different people. To a farmer a "good" day is related to how the machinery ran and the number of acres covered. For us a good day is being able to seed 150 acres.

Robbie seeded uneventfully all day. I could have taken the drills now that I'm done planting beets but I want to give him as much experience as possible. He's doing a good job and if I relieved him it would appear as if I were dissatisfied with his work. Besides, he can drive straighter than I can.

I cultivated Marion's west 80 a second time, attempting to tear out the curled dock. Hoping to seed into last year's stubble, I burned a hundred acres on Marion's south quarter. Forking the fire along, the gusty east wind carried the flames with a loud crackling. Within a few minutes the entire hundred acres was black with only a few pockets of smoldering straw among the charred debris. I drove my heel into the newly burned land and it sank almost to the ankle. With the straw removed, the field stands a much better chance of drying.

TUESDAY, MAY 27

6:00 a.m. Marion's south quarter. As I fixed a couple of fertilizer hoses on the drills, I noticed the complete tranquility of the morning. A meadowlark on the power line called his soprano song. Somewhere in the distance his mate would answer. The red-winged blackbirds clucked among the pond's cattails. Casting long westward shadows, the trees stood silently in the hushed wind. Everything was peaceful. Having spilled fertilizer on my hands, I walked to the nearby pond to wash. When I stooped to the water, a rat ran in front of me, jumped to a floating plank, missed and splashed into the water. Scrambling back onto the board, he ran to hide himself in a water's edge thicket. I looked at my watch. 6:20. Robbie will be here in forty minutes. As I resumed working on the drills, I felt the earth vibrate and heard the blast of the morning freight train fill the quiet air. Emerging from the north end of Kennedy with its big diesel engine growling, it signaled the commencement of the day's activities. Almost instantaneously a neighbor switched on his aeration fans. The sound of air being forced through a bin full of wheat could be heard a half-mile away. Semi-trucks suddenly appeared. Upshifting while accelerating, their engines poured black exhaust through overburdened mufflers. Cars and pickups also materialized on the highway. The meadowlarks and blackbirds could no longer be heard. Gone was the tranquil world I had known just a few moments before. Filled with anger at how such a beautiful morning was being ruined, I quickly realized that I was no better. Climbing into the cab of the Versatile, I pressed the starter button and the diesel engine came to life with a roar.

The "Things to Do" list hanging on the shop wall was especially ambitious this morning. Through the use of daily and weekly activity lists, I find I can accomplish more jobs in less time. Although lists can be quite beneficial acting as a work stimulus, they can also add undue pressure. Angrily the list stared down at me while I frantically worked for its appeasement. A ruthless taskmaster, it seemed to shout, "Do this!" "Do that!" "Better hurry!" "Why haven't you finished?"

Greg Maloney, a local dealer, brought fuel.

"Mr. Carlson, how are things going?" he asked with a smile.

"Okay," I answered, not looking up from my job.

"Been a nice spring, hasn't it?"

"Yeah," I said, walking away from him to get another tool.

Undaunted, he persisted. "Got most of the crop in?"

"Most of it."

"Have any combining to do this spring?"

"About fifty acres."

Poor guy. He was trying his best to be friendly, but my unsociable attitude finally got the best of him and he excused himself to watch the fuel tank fill. After he left I felt embarrassed by my rudeness.

Angry at my list, I looked up at it and said, "Look what you've done."

The last few days have been unseasonably hot for May. Just what we want. Having fifty acres of wheat still lying in the swath from last fall, I make daily inspections in hopes the ground will be firm enough to carry the combine.

In the afternoon I sprayed twenty-five acres with 2,4-D for curled dock. Wheat is seeded on this particular field but has not yet emerged. I've never seen such a heavy curled dock infestation as this year.

Cenex spread more urea today and I spent the remainder of the day cultivating. Carol and the four kids brought supper to the field. It's good to be near the end of seeding.

WEDNESDAY, MAY 28

Another day in the 90's. Terrific drying weather. The cultivator was again digging the dark gray ground. Tearing and throwing the overwintered soil, it incorporated the white pellets of nitrogen fertilizer into the root zone. The finishing harrows laid a smooth seedbed in which the wheat will be planted. The hot sun quickly evaporated the surface moisture, transforming the black earth into hues of gray. A dry, one-inch layer of crumbly topsoil developed and beneath that, a horizon of rich, moist strata. Into this, the wheat seeds will be pressed.

I finished cultivating, and spent the rest of the day picking a few stones and after-harrowing. Normally we don't touch a field after it has been seeded. This year, however, because the ground is so uneven and rutted, some harrowing must be done to smooth the way for later passes by sprayer, swather and combine.

After-harrowing is a practice of smoothing a newly-seeded field. It is not as common as it once was, because post-emergence herbicides, which are sprayed after the wheat has emerged, are dependable and widely used. Ten years ago more pre-emergence wild oat and pigeon grass herbicides were used. They were applied by spraying onto a newly seeded field directly in front of a harrow. The harrow would lightly incorporate the herbicide into the soil. As the undesirable weeds would grow through this zone of chemical, they would die. Sporadic results with the pre-emergence herbicides have made them less popular.

Robbie continues to seed. After today he'll have only forty acres left.

THURSDAY, MAY 29

Early this morning Harvest States spread urea on our last forty acre piece to be seeded. It's the field I burned on Monday and I'm hoping the week's hot weather has dried it sufficiently. Supposedly, there's a urea shortage this spring. At least that's what the fertilizer people tell us. Many farmers have had their work delayed, waiting for fertilizer. Having purchased most of our fertilizer from two different companies, I've been playing one against the other. Whoever can come the quickest gets the job. In the hot days we've been having, we don't want the urea to lie too long before incorporation.

The thermometer at Harvest States Fertilizer Company read 96 degrees.

"Look at that," pointed one of the employees. "That's gotta be some kind of record."

Sitting in the cool air-conditioned office sipping a cold Pepsi, I felt relaxed, knowing I had only forty acres to seed. All was quiet. The conveyors, front-end loaders, blenders, floaters and trucks sat motionless. Harvest States' full-time employees, plus their seasonal help, were sitting with me. Conversation was rare, and what there was was directed toward the morning train and the anticipated carload of urea. The phone rang.

"Here we go again," muttered one of the employees.

"Where's the floater you said would come this morning?" laughed another.

Terry, the plant manager, picked up the receiver.

"Harvest States . . . No, not yet . . . We're expecting a car this morning . . . Oh, maybe around 11:00 or so, depending on if the darn train is on time or not." Nervously fidgeting with his pen and doodling on a writing pad, he looked at us and rolled his eyes. His desk was piled with papers. A wall full of maps and lists of fields yet to be spread indicated a heavy work load in the coming days. With shades pulled, the room was dark and blue with cigarette smoke. A fat, lazy dog slept by the counter. The floor, not washed all spring, was filthy with caked dirt. Parts and broken pieces of fertilizer application equipment were scattered on the floor. The pop machine hummed in the corner. A half-full box of Harvest States caps sat on the counter. Staring out the window at the passing cars, one of the hired men puffed lackadaisically on a cigarette. Except for Terry's voice, the room was silent.

"Well, we're doing the best we can," he said in an apologetic tone of voice. "But until we get some product, our hands are tied, too."

"Too bad more anhydrous couldn't have been put on last fall," commented Steve Holmgren, drawing on his pipe. A manager with Harvest States, he was well aware of the logistics of transporting fertilizer from the plant to the field.

"Yeah," added another, "that's what screwed this spring up so much."

"I suppose the price of urea is going up too," I venture.

"You bet," Steve answered. "Whenever there's a big demand you can bet what the price is going to do. This car will be quite a bit more than the others. Not our fault. They're charging us more for the stuff so we've got to do the same."

"Another irate farmer?" I knowingly nodded in the direction of Terry who was by now squeezing and dismantling his pen.

"If we're lucky, maybe around 3:00 this afternoon," Terry answered into the receiver.

"He does well under pressure," I laughed.

Still staring out the window, the man smoking a cigarette added, "You'd be surprised to know how many farmers are the nicest people in the world except at seeding time when they are constantly hounding you, 'When are you going to come out and spread for me?' "

"Some can get downright nasty," contributed another, munching on a ham sandwich.

"Now, now, boys," Steve interjected. "It isn't all that bad." He looked at me and laughed. But it can get a little hectic around here at times." Terry put down the phone.

"The usual," he muttered.

We all nodded.

Robbie incorporated the freshly-spread urea on Marion's this afternoon. Both the floater and the 850 made some pretty severe ruts, but we are hoping that a rain will balance the fact that the ground was worked too wet.

This afternoon I walked among the wheat swaths. Shedding their blanket of white snow, they have emerged ugly and stained. The dry straw cracked as I tore it between my hands. The discolored seed snapped between my teeth. Tomorrow I'll try to combine. As much as I'm appreciative of the hot, dry spring, a thought appears in the back of my mind and grows daily. You hear it in conversations everywhere:

"The dry weather is great, but we could use some rain."

"It's too early to be this hot."

"We're in for a long, dry summer."

"Reminds me of 1980."

FRIDAY, MAY 30

Robbie seeded the remaining forty acres of wheat today. It's a relief to have the crop in. The temperature was over 90 degrees again today with a south wind. We could use a shot of rain.

Today I did something I've never done before in my life, and I hope never to do again. I combined wheat in the spring of the year. Actually, I couldn't have hoped for a nicer day. The stiff south wind thoroughly scattered the parched straw over the landscape. A neighbor stopped to visit. We joked about combining in May.

"Look there," he said. "When you look across the country, it looks like summer, spring and fall, all in one scene."

Indeed he was right. Across the coulee from where I was combining, our early-seeded barley was now getting some height. To the north a neighbor was seeding wheat. The three fields represented quite a contrast.

Surprisingly enough the field was firm, except for the far west side where I had to back out of a hole, leaving big ruts. At 8.5% moisture, it was the driest of 1985's wheat crop. Throughout the afternoon people drove slowly past, taking a good look at this unusual spectacle. Some cars came to a complete stop, their passengers gaping out the windows. My picture was taken. Someone unknown to me now has a picture of my combine in his photo album. Once when I was at the far end a pickup stopped, its occupant climbed into my truck box, examined the wheat and drove off. I felt like posting a sign for all curious people to read: "This wheat is yielding 30 bushels, has 12.3% protein, is 8.5% moisture and weighs 53 pounds."

SATURDAY, MAY 31

Lying flat on my stomach, my face inches from the ground, I witnessed a life and death struggle. The southwest wind blew relentlessly across the face of the beet field. Dirt particles skipped across the surface, causing me to squint. My mouth had the taste of grit. The howling wind tugged at the power lines and bent the trees. The tiny, recently-emerged beet plants are in their most delicate stage. Seemingly attempting different tactics, the cruel wind alternately pulled and twisted on the cotyledons. The thin, thread-like stems bent and coiled under the strain. I cupped my hands over a small plant, its cotyledons a half-inch long, giving it a brief respite from its battle. For some the fight was lost. Succumbing at last, they lay flat against the ground, their stems weakened to the point where they could no longer stand. I staked the spot. Two days later I would come back and find these plants dead. Their fragile leaves, lacking moisture, would crumble at my touch.

This infestation of wild oats in the beets is the worst I've ever seen. They are so thick that in places the rows can scarcely be seen. I stop and brush away the top layer of dirt. Underneath a second flush is massing, ready to break the surface. A post-emergence wild oat herbicide will be used, the eighteen-dollars-an-acre a necessity in the effort to save the crop. But when? Do I spray now or wait a week to get all the wild oats? What if it rains? Spraying by air adds an additional three dollars per acre. If I delay spraying I run the risk of the older wild oats getting out of control, becoming tall enough to interfere with the thinning. If I spray now, a second application may be necessary. The thought of wild oats robbing valuable moisture and sunlight from my beets not only makes me angry, but gives me a queasy feeling in the pit of my stomach. My mind is made up: "As soon as this cursed wind dies down, I'm gonna spray these suckers."

Robbie spent the day running around the country with the drills, seeding potholes. Coupled with the dry wind, the 95-degree temperature continued to dry the ground. I hope it rains.

MONDAY, JUNE 2

Not many days ago I would watch the nightly television weather show, hoping to see high pressure areas. Now I'm wishing for lows. An east wind has been blowing all day, usually a good sign of rain.

The Volkswagen engine purred under the spray coupe's hood as it straddled the beet rows up and down the field. The coupe resembles an airplane body sitting on a triple landing gear with fifty-foot spray booms as wings. Spraying Poast herbicide and ten gallons of water per acre requires stopping every two rounds on a half-mile field. Traveling at eight mph in the open cockpit created a breeze that felt good against my face. Underneath the machine, the green carpet of wild oats received a fine mist of deadly post-emergence wild oat killer. I hope it works. Deciding to experiment, I sprayed the most heavily infested half of Freda's quarter. Maybe I can uproot the wild oats between the rows on the other half with the cultivator and spray a seven-inch band on top of the beets, saving twelve dollars per acre.

Robbie and I spent the afternoon cleaning around the buildings and mowing the grass. Now that seeding is completed, menial tasks such as cleaning the shop, locating scattered tools and vacuuming tractor cabs can be done.

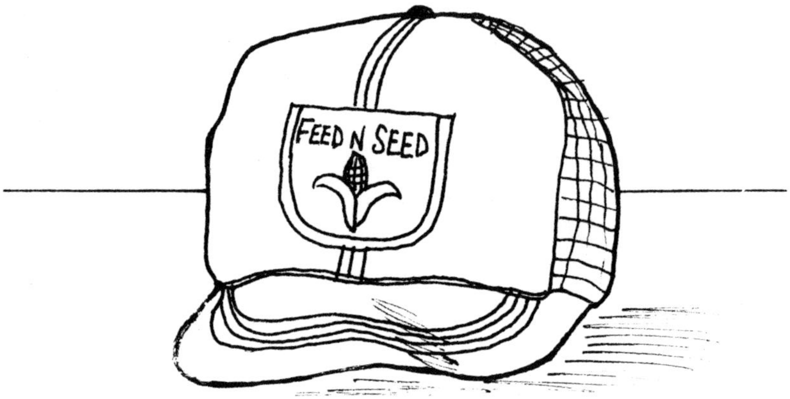

TUESDAY, JUNE 3

Yesterday's east wind proved to be an accurate omen. During the night an inch of badly-needed rain fell. The dry seed wheat lying in the wheel tracks will finally sprout. With the earlier germinating wheat already in the two-to-three-leaf stage, we'll have "two crops," one maturing several days before the other, creating harvesting problems. Most of the wheat will have to be swathed rather than straight combined. Nature isn't always fair. Four miles north, the rain suddenly subsided, dropping only .15 inch. Two-thirds of our crop now has adequate moisture and the potential of yielding well. The other one-third will have irrecoverable drought damage.

We almost had to shout to be heard above the loud baritone roar of over a hundred farmers in the metal building. One of the local fertilizer and chemical companies in Hallock was giving its annual steak fry. Pallets of Hoelon, Avenge, Buctril and 2,4-D ester rose to the ceiling. In the middle of the warehouse sat tables of beans and potato salad. We waited in line while steaks fried on a large grill next to the open door.

"Looks like a good crowd," I commented.

"Just give a free steak and you'll get every farmer in the country," smiled a farmer behind me.

He was right. Arriving in pickups, they clustered in groups, clutching beer or soft drinks served on ice in plastic grain hoppers. Some came from the field, especially from north of Hallock where the rain missed. Blue jeans with light blue cotton shirts or khaki work clothes dominate the attire. Old multicolored dress shirts, now relegated to becoming dirty and greasy, sprinkle the crowd. Caps advertise local elevators, fertilizer companies, chemicals, seed varieties and machinery dealerships. The evening is warm and humid. One-liner jokes are volleyed among the crowd. The air is friendly and relaxed. People are interested in knowing what others are doing. Conversations open by: "How much rain did you get?" "How's the crop?" "Done any spraying?" "Lose any beets in the wind?"

Sipping a cup of coffee, I sit on a five-gallon can of Buctril and discuss the current crop conditions with a neighbor. To my right, "How the U.S. Government Is Manipulating the Wheat Market" is the subject of a dogmatic monologue being preached to anyone who cares to listen. Gesticulating with an Old Milwaukee beer can, the orator unfolds his dramatic conspiracy. A jocular crowd to my left banters about the pros and cons of green versus red machinery.

Later, driving home and smelling of charcoal, I considered the evening a success for both the sponsoring chemical company and me. No doubt the steak-fry will sell a lot of chemical, and I found out what has been occupying other people's time, what seed varieties are best, how much rain everyone received, why the Versatile 900 will outpull the John Deere 8630, what the best beet seed spacing is and why the CIA confiscated the satellite pictures of the Soviet wheat crop west of the Volga River.

WEDNESDAY, JUNE 4

The humidity eased and the weather cleared. Twisted shovels were mounted on the chisel plow, and Robbie began working the recently-combined wheat stubble. I spent the remainder of the day recalibrating the beet band sprayer, and attaching the cut-away discs and side knives to the beet cultivator.

Making one of his routine calls, the American Crystal Sugar field man, Gary Moe, and I walked one of the beet fields. He had a method of counting the number of beet plants in a three foot stride, moving to the next row and counting again. I would step ten paces down one row, approximately thirty-three feet, and count the plants. The object is to ascertain how many plants there are in one hundred feet of row.

"I get around two hundred plants," I said.

"At least," he agreed. Lighting a Marlboro, he facetiously continued, "I hate to tell you this, but you've got too many beets."

"That's the kind of problem I like to have," I laughed.

"You're right there. A man can always thin them out."

Tonight the National Basketball Association's championship series is on television, and Melissa, Alan, Todd and I watch our favorite team, the Boston Celtics. Propping a milk bottle in Andrew's mouth to keep him occupied, we shout and cheer when Kevin McHale grabs a rebound and Larry Bird hits a long shot. At half time we race to the back patio to shoot baskets. After a rousing game of two-on-two in which we pretend that we're the NBA stars, we retreat to the basement to watch the second half. In the third quarter a fight breaks out between the Houston Rockets' seven-foot tall forward Ralph Sampson and a smaller Boston guard. The kids scream with excitement as Sampson gets kicked out of the game. Carol, who has been at a night class in Kennedy, arrives home at this time and wonders what all the noise is about. Melissa and Alan run upstairs shouting, "Mom! You won't believe it! This Boston guy just gave Sampson a little nudge and Sampson got mad and hit him and then they all started to fight and the referee had to break them up and all the players were on the floor and the crowd was just yelling and they kicked Sampson out of the game!"

"That sounds exciting," Carol said, "but who is Sampson?"

THURSDAY, JUNE 5

Robbie attached the field cultivator to the 850 and began working the summer fallow acres while I busied myself changing oil in the tractors.

The grain crop is now obscuring the dark ground in which it was planted. From now until July 1 there'll be plenty to do. In late springs such as this, the beet work, summer fallow cultivation and wheat herbicide spraying all run together without much spare time.

When I look at the wheat from the pickup window it appears weed-free. However, after just a few steps into each field, I discover an infestation of wild oats, wild buckwheat and mustard. Spraying will have to start shortly.

Today, Carol and I are celebrating our tenth wedding anniversary. Ten years ago today was a hot windy day. We were looking for a rain. The first year we were married Carol taught kindergarten in Kennedy and since then she has been my partner in the farming operation: running errands while I'm tied up on the tractor, bringing lunch and supper to the field, relaying messages from the CB to the telephone, helping with the book work, and most important, raising our four kids when I'm not around.

Tonight we took another couple, Jim Johnson and his wife Virginia, for supper in East Grand Forks. Gorging ourselves on prime rib, we could look out the windows of the supper club and see a nice rain falling. On the trip home I kept hoping that the rain had continued to the north. Arriving in Kennedy, however, we discovered that we had only received a trace. It was disappointing.

SATURDAY, JUNE 7

What seemed like a couple-hour's job at the onset became a full morning's work, and could have lasted all afternoon if we wanted. Four isolated spots of curled dock were growing in the beets on Freda's quarter. Appearing like distant islands, the tall, thick bushy plants have been an eyesore and a source of irritation as well as a hindrance to cultivating. Robbie and I set upon the task of pulling them by hand and throwing them into the pickup box. After pulling for a few minutes, it occurred to me that there were not only more dock than anticipated, but pulling the taprooted plants proved to be more strenuous than we thought. Determined to remain, each plant required a struggle before our leg, back and arm muscles prevailed and the weed exploded from the surface, leaving a large crater. Whether or not we were doing any good is a matter of debate, because whenever a curled dock plant was pulled, it would invariably dislodge any small beet plants in its vicinity. Exhausted after loading the pickup three times, we quit. For cosmetic reasons, the few remaining dock could be pulled, but my aching back and legs told me it wasn't worth it. Robbie went home and I shot a round of golf.

MONDAY, JUNE 9

Sugar beets and rocks do not mix. Even rocks as small as golf balls can cause problems. They get caught in the cultivator and thinner, thrown through the rotobeeter and stop the lifter's wheels and grab rollers. A couple of high school boys were hired through the local youth job service to remove these troublesome Ice Age remnants. Armed with five-gallon pails, they spent the day filling the box of the old pickup. When the springs could no longer hold the load, the stones were thrown into the ditch at the edge of the field where a new access crossing would be built.

What I consider the most tedious operation in raising beets started today. Carefully dropping the cultivator on the ground, I inched my way forward at two mph. The beet cultivator is equipped with cut-away discs that slice a thin five-inch band in which the tiny beet plants sit. Behind the cut-aways are long tunnel shields protecting the beets from becoming covered by dirt. Between the rows, long side knives blacken the ground. Shaving the ground an inch below the surface, the horizontal knives neatly remove any small weeds. Like miniature strip farming, each pass with the cultivator leaves twelve green five-inch bands separated by seventeen inches of rich, black Red River Valley topsoil. Slowly the tractor and cultivator grope their way down the field. With not much room for error I would occasionally get off the row, suddenly destroying twelve rows of beets. Clutching immediately, I raise the cultivator, re-align the tractor and start again. Physically the job is not taxing, but by day's end I'm mentally exhausted. The summer sun is over the western horizon when I drive slowly home.

TUESDAY, JUNE 10

The John Deere was halfway down the field when my watch showed 6:00 a.m. Back and forth I guided the tractor, being ever so careful not to tear out the beets. To pass the time I calculate how many rounds are being made in an hour and attempt to project where I'll be at noon, 6:00 p.m. and quitting time. In the distance, at the other side of the quarter, Robbie and the two young rock pickers are toiling in the sun.

Some of the most enjoyable meals of my life have been eaten sitting on a ditch edge on a warm summer day. Today's dinner has to rank in the top ten. Placing my opened lunch kit on a grass-covered gopher pile, I prepared a table for one. Nestling comfortably, I dined on peanut butter and jelly sandwiches, assorted cookies, a pear and lukewarm coffee. The 135 horses under the tractor's hood rested silently. A striped gopher nervously watched from a distance. A mile west a neighbor's tractor sat quietly, temporarily abandoned by its owner for the noon hour's sociability of a crowded restaurant. As I lay spread-eagled on a bed of bromegrass, thoroughly enjoying the serenity, the warm summer sun soon lulled me to sleep. All was peaceful.

Suddenly I awoke to the sound of stones scattering in front of radial tires and the roar of a Chevy Blazer's engine. In an instant he was upon me. Frightened, I sat up like a shot, my cap falling in time to see the dust trail of the mailman heading east. My watch read 12:25. Looks like lunch break is over. Groggily, with lunch kit in hand, I walk to the waiting tractor. At my command it comes to life. We spend the rest of the day creeping up and down the field. By evening, half the quarter will be cultivated.

WEDNESDAY, JUNE 11

After picking rock for two days, Robbie is glad to cultivate summer fallow again. He never complains, but I'm sure he must be a bit stiff. The weather is beautiful. Carol brought dinner to the beet field where I'm cultivating. Three blond heads scramble from the car shouting and waving. A fourth, Andrew, is in his car seat, sucking on a bottle of milk. After lunch the two oldest boys, Alan and Todd, want a ride. The six- and four-year-olds climb the steps and snuggle themselves to the left of the tractor's seat. Melissa, age eight, waits at the field's edge for our return. Easing out the clutch, we make the first round of the afternoon chitchatting about kittens, swing sets, climbing snow banks and Alan's attending kindergarten this fall.

Suddenly Alan screamed, "Look!" pointing to a couple of fist-sized rocks that the pickers had missed. "Who's been throwing rocks on your field?"

"God did," I replied.

"God did?"

Arriving at the end of the row, I raised the three-point, swung the tractor 180 degrees, found the next proper row and dropped the cultivator back in the ground.

"How did you do that?" asked Todd, looking back at the cultivator.

Brushing aside explaining the principles of hydraulics, I simply told them about moving a certain lever back and forth. Satisfied, they spent the rest of the round discussing frogs, bird nests and dinosaurs. Deciding that cultivating beets wasn't very exciting, they wanted to leave after one round. The once-muted engine roared with the opening of the cab door.

Wincing in the noise, Alan shouted, "Know what I'm gonna be when I grow up?"

"What?" I asked.

"A farmer."

"Me too," added Todd.

Descending to the ground, they ran in the direction of the car.

"So long farmers!" I shouted.

They turned their heads and grinned.

I've tried to keep my eyes and ears open to acquire more acreage as farmers in the area quit or retire. Fifteen years from now, if my boys want to farm, maybe they can be the ones to buy land if I can help them out by furnishing machinery or co-signing the note. It seems as if it takes a generation to get fully into farming when starting from dollar one.

I would like to purchase more land in the future. There have been times when I've bid on land but the deals haven't gone through. Maybe a man's better off in the short run. If I can continue to rent land for $40 to $45 an acre, it cash-flows better than paying $80 to $100 an acre in land payments for the next thirty years. But there's more security in owning land.

By late afternoon I reach the other side of the quarter. I slouch in the seat with my feet up, admiring my work. The time for self-gratification cannot last very long, however, as there are still beets by Hallock to cultivate. At dusk they too are done. It is 9:30 before the 4430 sits silently in the yard. Wearily I walk to the house for supper.

THURSDAY, JUNE 12

The wheat kernel appeared sound. Not finding any sign of sprout, I bit into it. The seed snapped between my teeth, exposing the white starchy endosperm.

"Maybe the seed is still good," I said to myself.

Digging further, brushing away the dry dirt, I found a few more. Most appeared viable. However, on some the embryo had cracked open with the small seminal root protruding. Running out of moisture, the seed died. I surveyed the wheat field before me. Two distinct sets of tracks where the seed had not germinated could be seen. One, at an angle, represented the cultivator. The other, perpendicular to the road, indicated where the drills had traveled back and forth. Like black lines on a green carpet, they formed one-hundred acres of twenty-foot parallelograms.

Amidst the wheat the wild oats are mostly in the two-leaf stage. If Hoelon, a wild oat herbicide, is to be sprayed, next week would be the optimum time. But is the crop worth spraying? If the weather stays dry and we eventually lose some crop, why invest any more money in a lost cause? If it takes six bushels of wheat to pay for the cost of spraying, are the wild oats decreasing the yield by that much? These and other agronomic and economic questions mull in my mind as I walk a barren drill track to the road.

The Ford pickup slowed to a stop. Its occupant, who farms a couple of miles down the road, hung both elbows out the window.

"Funny how it can't seem to rain," he lamented.

"I'm afraid we're going to have some pretty poor crop," I conceded.

Stepping out, he pulled a bromegrass stalk from the ditch. Inserting it into his mouth, he pointed to my field.

"Will the tracks come with a rain?" he asked.

"I think most would, but some have germed and died," I answered.

Leaning against the pickup, he cast an experienced eye over my field. "It's going to be a mess when they do come. This has happened before. Looks like a guy's gonna have to let it stand as long as possible before swathing. Personally, I hope the tracks don't come. They're not going to amount to a hill of beans anyway."

"The swaths will have to lie a long time to cure," I nodded.

"Well, sure," he agreed. "That just increases the odds of getting rained on." Seemingly embarrassed by his outburst, he smiled. "Oh, well. A guy just has to do the best he can."

Little sugar beet plants are very poor competitors with weeds. The Poast is doing a terrific job on the wild oats, but because of a second flush, I band sprayed sixty acres a second time. It was my own fault. If I had trusted the chemical more and waited before spraying the first time, I could have killed all the wild oats. The mistake cost me an additional seven dollars an acre. Later in the afternoon I switched to another herbicide used in sugar beets, Betamix, which is designed to kill small wild buckwheat and other weeds. I sprayed until a rain shower forced me to quit.

FRIDAY, JUNE 13

Referring to the .1 inch of rain received yesterday afternoon, a non-farmer remarked to me how every little bit helps. Letting the comment slip without response, I thought of how wrong he was. Unless enough rain is received so that its precipitation meets the subsoil moisture, it will do no good at all; it merely wets the soil's surface without soaking in to meet the wheat seed. By late morning the summer sun had already evaporated the meager life-giving fluid.

The weed leaves were dry and I began band spraying. Spraying only a thin band over the top of the beet row cuts the herbicide cost by a third. Each succeeding operation with the beets goes a little faster. Cruising down the rows at five mph seems a blinding speed compared to the first cultivation. Pumping the equivalent of thirty gallons of water per acre, the band sprayer was actually discoloring a seven-inch band over the beet row. Looking back, I saw twelve dark gray stripes momentarily being painted on a light background. But they vanished before my eyes as the warm air quickly dissolved my handiwork, and the gray and black earth tones blended into one color. The day passed quickly. I stopped every twenty acres to refill water and herbicide, and the acres rapidly were finished.

The beets are looking very good. Too bad I can't say the same about some of the wheat. Where the rain missed, the grain crop will be quite meager. Even if a rain does come, the yield is permanently retarded to the point where I deem it useless to incur further expense in herbicides. Recognizing a lost cause, I consult my crop insurance policy, which tells me that there will be more dollars-per-acre income collecting insurance than pouring more expenditures into a pitiful crop.

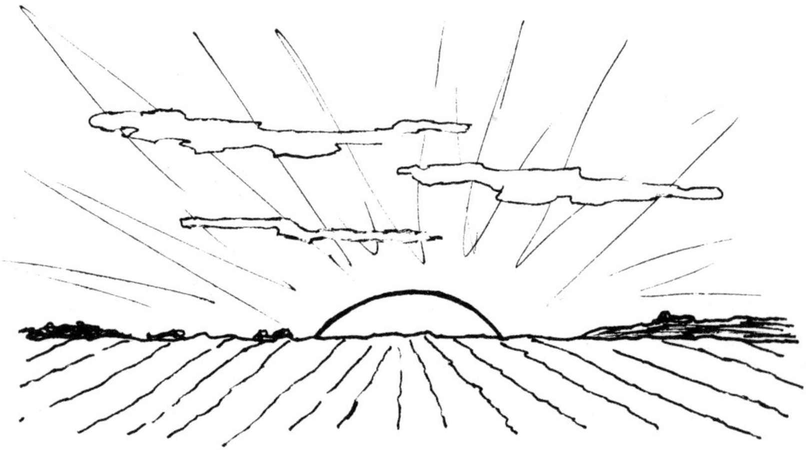

MONDAY, JUNE 16

Despite the fact that the sun was but a few degrees over the eastern horizon, the morning was warm. The tall cottonwood tree row dividing the quarter in half stood silent. A killdeer viciously scolded me as I stepped from the ditch to plod along dark green rows of hard red spring wheat. This particular eighty-acre field on Section ll, fortunate to have received rain, is looking excellent. There is no dew and my boots remain dry. A sprinkling of three-leaf wild oats can be seen, and I decide to spray.

Later that morning while standing at the parts counter buying a modulator for the beet thinner, I heard a familiar voice behind me.

"When are you going to thin those beets?"

Recognizing Jamie Peterson's* voice, I braced myself for the inevitable lecture.

"This afternoon," I defensively answered. As if it were any of his business.

"It's about time. They could get away on you."

"I don't think so," I smiled.

"You don't think so! One rain and you're screwed. Then you'll be in there chopping the heck out of them. That's why I space plant. Put 'em out to five-and-a-half inches and just harrow once or twice if they're too thick. It's the only way to go. Even sold my thinner a couple of years ago. Just sat in the Quonset never being used. Saves a lot on the seed bill too."

Jamie was a likable sort of guy but he had one of the most opinionated and argumentative personalities around. Electing not to argue, I mentally pleaded for my part to be found. Ken Urbaniak, the owner of the dealership, had relieved the parts man for the noon hour. He had just spent the last several minutes searching the parts bins for what I needed.

"Found it," he called. Blushing slightly, he laid a three-inch black rectangular plastic object with electrical prongs at one end on the counter. "I had my numbers mixed up."

Fondling my newest purchase, I mistakenly asked, "What's it worth?"

"I was afraid you'd ask that," Ken replied, filling out the purchase slip. "Two hundred fifty dollars."

"There's another reason I don't like to thin," Jamie said.

All I could do was shake my head and laugh as I quickly excused myself.

The thinning process is perhaps one of the most critical operations in raising beets. If the beets are thinned too thick, they will not grow as large and tonnage will be lost. If the beets are thinned too thin, tonnage will also suffer, accompanied by a decrease in the sugar content. There are two schools of thought on how to arrive at the proper plant population per acre. Perhaps the one with the most popular appeal is to space the seed farther apart at planting. One of the most vocal advocates of this, of course, is Jamie Peterson. All the advantages of space planting were mentioned this morning in Jamie's dissertation. They are savings in labor, fuel, seed cost, repairs and machinery. A notable disadvantage is that if the beets are planted thin to begin with, and Mother Nature decides to do more thinning of her own through frost or wind, a much thinner-than-desirable stand can result. The other method is to plant thick, and thin either manually or mechanically. The costs are higher but if it means the difference between a good or poor stand, these costs can be quickly recovered in increased yield and sugar content. In our rough seedbed this spring I opted to plant heavily.

The John Deere 200 series beet thinner is a marvelous piece of equipment. Its own generator produces 110 volt electricity and an electrical circuit is completed when a probe touches a beet plant. This trips a hammer to slice across the row, tearing out unwanted beets. Plant spacing can be adjusted by changing the length of the cutting knife attached to the hammer, by adjusting a delay switch to order the hammer to trip rapidly or slowly, or by varying the tractor ground speed.

I spent the afternoon working with these three variables to achieve a plant spacing of 120 to 140 plants per 100 feet of row. The ground is dry and firm. The thinner throws dirt to the left and right as the hammers alternately slam back and forth, chopping out the unwanted beets. Leaving a twelve-row path of destruction, they ruthlessly uproot and dismember the plants. From the tractor cab it appears as if the field is being ruined. Many times I find myself stepping off ten paces and counting the plants remaining behind the thinner. The beets are battered and chewed apart. Raking my fingers over the freshly-thinned row, I remove the carcasses of the uprooted beets. Even those that are left look as if they'll soon die of the terrible beating they've just received. Some are barely holding onto the ground by the end of their roots. Others have had most of their leaves chopped off. I tell myself how foolish it is to treat a perfectly healthy crop this way. However, past experience tells me that within a week they will fully recover. I finished the beets by Hallock and drove the tractor home in the dark.

TUESDAY, JUNE 17

After sitting on the edge of my chair for half an hour, I was relieved when the elevator board meeting was finally over at 11:00. I raced home, grabbed my lunch kit and drove the tractor to the beet field for another day of thinning. The day was warm with a slight westerly breeze. The thinner is working well, and in spite of my late start I'll cover quite a few acres this afternoon.

"What a terrific day," the radio DJ proclaimed. "Take my advice and get to the nearest swimming pool."

Remembering that recommendation I find myself wishing I were on the golf course or taking a swim. Monotonously the afternoon droned on, when suddenly I saw a young meadowlark stumbling into the thinner's path. Clutching, I allowed him to clumsily half-fly, half-fall to safety before proceeding. I'm glad I saw him in time. The rest of the day is spent riding up and down the rows. The beets are uniform and I don't have to get out much to count the stand. I relish the thought that I'm doing as much in one day by machine as eighty good laborers, without paying any workman's compensation, social security or having to speak Spanish. By 9:00, eighty acres are done. The tractor idles for a few minutes before being shut off. The beets that are left are a little small and it's not as critical to thin them immediately. The weather forecast is continued warm and sunny.

Enjoying the cool of the evening on the patio with a cold glass of iced tea, I quote Ecclesiastes to myself, "There is nothing better for a man than to eat and drink and tell himself that his labor is good."

WEDNESDAY, JUNE 18

Another forty acres were thinned by noon. It was looking like a terrific day, until shortly after dinner, when my once-dependable thinner started to develop problems. Noticing that one hammer was tripping continuously, taking out all the beets, I stopped immediately. The most obvious cause was a clogged probe. When I found that it was okay, I experimented with various remedies: switching modules, removing and licking fuses, removing components from the seemingly faulty row to a good unit. It still didn't work. Well, when in doubt, read the operator's manual.

The title on the yellow and green cover was: *Operator's Manual, John Deere 200 Synchronous Thinner OM-N159342 Issue L3.* Sounds official. Let's see . . . "Trouble Shooting," pages 22-24. Under the column, "Knife oscillates continuously and uniformly," I found four possible causes followed by four remedies. Seems simple. Half an hour later the apparent simplicity had vanished somewhere in the maze of circuits and wires. Then suddenly the knife seemed to operate properly. Driving down the field, I was relieved to have the problem solved, although I didn't know what I had done to solve it. A few minutes later I found myself standing behind the thinner, scratching my head. The problem was the same, except on a different unit. After much experimentation I discovered this one also miraculously started behaving as it should. Perplexed, I thinned another half-acre before a third unit began continuously chomping back and forth, removing every beet in its path. Jiggling wires and fuses seemed to be the needed remedy, but only for another three hundred yards before the problem recurred. By then I had memorized the operator's manual and decided to call John Deere for expert advice.

The white service pickup was not long in arriving. The two mechanics patiently listened to my description of symptoms before one asked, "How long has it been since the tractor's hydraulic filter has been changed?"

Thinking for a moment, I replied, "A hundred hours."

"Let's try changing it."

"It shouldn't clog in a hundred hours."

"We've seen it happen before."

A few minutes later the thinner was working perfectly. Embarrassed but happy, I was able to thin for the rest of the day.

THURSDAY, JUNE 19

All week while I've been thinning, Robbie has been spraying Hoelon. The wild oats are in the two-to-three leaf stage, and the optimum time to kill them is at hand. Agronomists tell us that under ideal conditions, a cereal plant such as wheat or wild oats will sprout additional leaves in four days, thus decreasing the wild oats' vulnerability to the herbicide. With that narrow a window of susceptibility, it's nice to spray while the weather permits.

The thinner was working properly and I began to hack away at the beets. The beets I worked on today are somewhat smaller and easier to thin. The stand, however, is a bit more erratic, resulting in numerous stand counts and adjustments. All day the knives neatly sliced a space between the plants, leaving a blocked appearance. Already the earlier-thinned beets have recovered to the point where one can hardly tell that they had been torn apart by the mechanical thinner. My biggest concern now is the carpet of wild oats growing between the rows where I had only band sprayed. The weather forecast is for a chance of rain. I hope it's correct.

The hours tick by and before long another beautiful summer day has escaped amid the roar of the tractor engine, the scream of the thinner's generator and the clomping of the trip hammers. By day's end only another morning's thinning remains.

When the tractor is shut down, my footsteps on a lonely township road are the only sounds I hear as I walk toward the waiting pickup in the gathering dusk.

FRIDAY, JUNE 20

"Hello?"

"Rob, take the day off. I'll call when it goes again."

"Okay."

Not one to waste words, Rob quickly hung up the phone—no doubt to go back to bed. I wouldn't blame him. He's been putting in some pretty long days and deserves a break. A healthy stream of water was pouring from the drain spout as the rain intensified.

"Looks good," I thought. "We can sure use the rain."

My hopes for an all-day soaker were soon dashed as the rain gradually diminished, and a band of blue sky appeared on the western horizon. By late morning the skies were clear.

Farmers are creatures of habit, and their actions can be predicted with astonishing accuracy. For instance, send a half-inch rain (such as this morning,) bring out the sun and you'll find them tracking every dirt road in the township, inspecting fields. I was no exception. After dinner, I found myself fishtailing down a seldom-used road. A freshly-dug pair of tracks indicated I was not the first to travel this slippery path. At the edge of a neighbor's beet field, a rain gauge had been set. Muddy footprints from road to gauge denoted a concerned farmer.

SATURDAY, JUNE 21

My birthday. If I had a government job I could take the day off with pay. I did, however, afford myself the luxury of sleeping until 8:00.

The cool morning warranted a jacket. Looking up from the welder, I saw a familiar pickup, uncharacteristically dirty, splashing through a muddy water puddle en route to my Quonset.

"Fixing, fixing," the neighbor said with a broad grin.

"Seems like it never ends," I answered.

"That was a nice little rain yesterday. We could have used more, but guess we'll take what we get and be satisfied."

I smiled in agreement with the traditional farmers' logic.

"How's the crop?" I asked.

"Good, where we got rain earlier. Some of the other stuff is going to be thin."

"Ours is the same way."

"Can you imagine what sort of people we'd be if we got bumper crops every year?" he philosophized. "We'd be so proud that no one could stand us."

"That's a healthy way of looking at things," I said.

"God has a terrific way of keeping us farmers humble. Just shut the rain off for a while or turn it on too long. You realize what an insignificant role we play in raising a crop."

Knowing his easy-going attitude toward life, I never doubted his sincerity. After several minutes of tranquil conversation he slowly drove down the road in the direction of his home. Watching him go, I was grateful for people who keep life's events in their proper perspective.

The longest day of the year brings with it the most lush and green time of the summer. The fragrance of the lilac bushes' purple blossoms can be scented through an open pickup window at thirty miles per hour. Lawns, ditches and crops are a healthy dark green. Ducks that nested in the ditch have disappeared. No doubt they've hatched their young and have led them to the river's safety. Regrettably this is also one of the busiest times. The grain and row crops demand attention and if one isn't careful, he can miss what nature is doing all around him.

A very gentle northwest breeze made conditions ideal for spraying 2,4-D ester on Marion's south quarter. A fine mist of 80:l ratio of water-to-herbicide from a small orifice at sixty pounds of pressure leaves an invisible coating of growth hormone. The wheat with its narrow, waxy, cuticled leaves will not be harmed. Conversely the broad, coarse, absorbent leaves of the unwanted weeds will soon take in enough of the chemical to grow themselves to death. Keeping the spray coupe aligned properly permitted only fleeting glances over my shoulders. The triangle-shaped spray patterns spewed from each nozzle merged just above plant height, blanketing the ground with a rolling mist. Penetrating the wheat's canopy, it bounced off the ground, briefly swirling before encircling and adhering to the stems and leaves of the wild mustard and wild buckwheat. Having a few wet holes, the field is somewhat soft in places, and the coupe's wheels manage to collect a little mud along with a few wheat leaves. I finished by late afternoon and towed the spray coupe home behind the water truck.

SUNDAY, JUNE 22

After church in the morning and an afternoon of swimming, Carol, the kids and I visit a neighboring church where I've been invited to present the evening's program. I have memorized the Gospel of Mark and enjoy delivering it in a theatrical fashion to churches and organizations. The entire sixteen chapters take two hours, so usually I do just half at a time. It's always a rewarding experience for me.

MONDAY, JUNE 23

This morning's work was a leisurely way to complete beet thinning. Parking the thinner in the pole shed, I savored my good fortune to have the thinning process accomplished under such ideal conditions.

Wanting to use the spray coupe for spraying wild oats in beets, Robbie thoroughly rinsed any remnants of 2,4-D ester from its tank and lines. Unfortunately, just then the wind rose to the point where spraying became impossible. It wouldn't be a normal year if the wind didn't blow, disrupting the spraying. The danger, of course, is that the particular weed you're after will grow too large to be controlled with herbicide.

It wasn't too windy to hinder band spraying. Because the nozzles are only seven inches above ground level and there are shields on either side, it takes a pretty stiff wind to disrupt the spray pattern. A few small wild buckwheat were growing among the beets and they could be taken out with Betamix. Robbie spent the afternoon steering the 4020 up and down the beet rows, band spraying.

Sitting in Kittson Auto and Implement (Hallock's Versatile dealership) on what appeared to be a church pew shortened into a love seat, I watched as a farmer nervously fidgeted with his newly-purchased sprayer nozzles. Examining each one individually as if to discover a major flaw, he finally stated, "These'll do," and signed the farm plan. "Farm plan" is a way in which the implement dealers work with the local bank. All charges are paid to the implement company by the bank. The bank then carries the account. If payment is late, interest will be added. For its services, the bank charges the dealers four percent of the amount of the purchases. The dealers are willing to pay the four percent because it assures no bad accounts and timely payments.

The afternoon was hot with a gusty south wind blowing through the open doors. Above my head pamphlets advertising new farm machinery gently flapped in the breeze. Under my seat, the pan of mouse poison appeared to have been nibbled on. Hanging from the ceiling, the flypaper revealed a bountiful harvest. Hundreds of diverse items such as hydraulic cylinders, roller chains, grain auger flighting, toy Versatile tractors, light bulbs and cultivator shovels hung on the walls.

Turning in my direction, the farmer-customer said, "Too windy to spray."

"Yeah," I replied, not knowing if it was a statement or a question. "Maybe it will go down by evening."

"I hope so. We've got a lot to spray. If I'd have known it was gonna blow, I'd have sprayed earlier." He paced the floor and climbed on a garden tractor. "I should have sprayed the other day," he continued, "but I thought I'd wait for the wild oats to get a little bigger. If this keeps up," nodding to the open door, "I'll have to switch to Avenge." Avenge is another post-emergence wild oat chemical which can be sprayed when the wild oats are in the five-leaf stage.

This farmer was always second-guessing himself, and his sentences were richly sprinkled with "I should have's." I too was concerned about spraying, but tried not to betray my feelings. Rational thinking kept telling me that in previous years the wind has always subsided, but unfortunately emotionalism began to gain the upper hand. Driving home and thinking the worst, I imagined sod-bound wild oats choking the wheat and sugar beets of all moisture and nutrients, completely suffocating the crop.

All afternoon I waited, but the wind continued to blow. Finally by early evening it diminished, and by 8:00 I was spraying Poast in the beets. The picture of health, wild oats with their dark green leaves represent the greatest challenge to the beets. I hope the chemical works. The western horizon still had tints of twilight at 10:35 as I drove the water truck home. Anxious for an early start, I'll be spraying before sunrise tomorrow morning.

TUESDAY, JUNE 24

A knock at the door awakened me at 7:00.

"Oh no," I groaned.

Sure enough, Robbie was at the door. Bolting from bed, I ran downstairs in my underwear.

Standing outside the screen door, he said with a smile, "Looks like you're not quite ready to go spraying. Want me to go out and start?"

"That sounds like a great idea," I moaned. The weather vane atop the pole shed pointed straight south. "Wind's picking up pretty good from the south."

"Forty percent chance of rain."

"You go ahead. I'll be out in a few minutes."

Of all mornings, why did I have to oversleep today? I skipped breakfast, quickly dressed and ran out the door. Arriving at Freda's quarter, I saw Robbie was already on the far side, making his first round. I watched as he swung the spray coupe 180 degrees and began the half-mile trip in my direction. Studying his spray pattern carefully, I watched intently as an occasional gust would disrupt the milky-colored mist. Reaching the field's end and shutting the gate valve, Robbie stopped the spray suddenly as he pivoted the three-wheeled sprayer. Staring at the boom's end for proper alignment on the next round, he opened the valve and gave the German engine full throttle as another fifty-foot swath was covered. Conditions are borderline. If the wind rises further we'll have to quit. To save time, I busy myself premixing chemical and spotting the water truck. As luck would have it, the wind diminishes and we are able to finish the beets and switch to grain. All day, acting like a racing team, we refill the 130-gallon tank with water and chemical. By day's end we've sprayed many acres.

WEDNESDAY, JUNE 25

Robbie finished spraying Hoelon on Marion's north quarter. Moving to Dad's, he started spraying 2,4-D ester, knocking down the broadleaf weeds. Most of the wheat is now in the five-leaf stage and can be sprayed with a stronger herbicide.

The beets have really grown in the past couple of weeks. Another ten days and they will be filling in the rows. Although they were planted three weeks later than desired, the warm weather has been ideal for them to catch up somewhat. Cultivating beets will occupy my time today. The second cultivation involves less danger to the now-larger plants. Since less accuracy in steering is required, the ground speed is increased to four mph. Surprisingly clean, the fields appear to need only two cultivations this year. Up and down the field the tractor moves, leaving behind freshly tilled soil between the rows. A devout worshiper of the sun, I appreciate this kind of day. Hot. The hours pass slowly. Self-pity takes hold of me. What am I doing here? I should be swimming or golfing or lying on a blanket enjoying the day instead of being out here. The radio announcers don't help either.

Grand Forks: "Hey, it's a beautiful day. I hope you're enjoying the beach. Well, splash a little suntan lotion on for me and stick with us as I play your favorite hits."

Fargo: "If you like summer, you're going to love today. Do yourself a favor and head to the park and catch some rays."

Winnipeg: "To all of you on Grand Beach, it's time to roll over and tan the back as well. So dig your toes in the sand and enjoy."

Self-pity has a new companion. Guilt. My kids would love to go for a swim but I'm too busy to take them. Some father I am. They'll probably grow up remembering how Dad was always too busy to take them anyplace. Finally I could stand it no longer. Parking the tractor, I raced for home, picked up the three oldest kids and took a refreshing swim at the Hallock pool. Carol stayed home with Andrew, just six months old. Melissa, Todd and Alan were really appreciative, and after two hours of swimming we are at home, sipping iced tea on the back patio.

Feeling refreshed and with my self-pity and guilt stymied, I cultivate beets with a renewed sense of enthusiasm. The hours pass quickly until at darkness the last round is made. I smile to myself, thinking of the good day it has been.

THURSDAY, JUNE 26

"Mike, do you have a copy?"

"Yeah, what's up?" he answered over his CB.

A half-mile away his shiny 4250 John Deere reflected in the morning sun.

"Let's go golfing this afternoon."

"Naw, I better not. Gotta finish this field today."

Try as I might, I couldn't persuade him to leave his beet cultivator.

Fun and golf are the order of the day as the Drayton beet plant's annual golf stag is held at Stephen. The break is good for everyone's spirits, and a couple of rounds of golf and a good steak supper do wonders. It's a day when farmers shed their seed caps and work clothes for golfing attire. Most are amateurs and scores are high. Conversations are light and fun-loving; but as diversified as the subjects are, the one that keeps coming back is sugar beets. The day is hot and humid. At 7:30, a dark cloud bank approaches in the northwest, appearing heavily-laden with moisture. We hopefully watch as an impending rain is almost assured, but our hopes are dashed as only a small shower develops and clear sky appears.

Beginning to show signs of drought stress, some of the crop has a lighter hue of green. We could use a good soaker soon.

FRIDAY, JUNE 27

Every day it appears as if the sprayed wild oats are turning color. A lighter green was first detected, and now they're actually turning brown. Some were getting pretty tall when they were sprayed, but the Poast seems to be taking them. I don't mind spending the money on herbicides when the results are good. What's frustrating is when adverse weather conditions, application timeliness or weed development result in poor weed control.

When I was pouring the chemical into the sprayer, the skull and crossbones caught my attention. Next to it, in two languages were the words "POISON," "KEEP OUT OF REACH OF CHILDREN," "DANGER." My curiosity aroused, I read further. "HAZARDOUS TO HUMANS." That's me. "FATAL IF ABSORBED THROUGH SKIN. MAY BE FATAL IF SWALLOWED. HARMFUL IF INHALED. CORROSIVE. CAUSES IRREVERSIBLE EYE DAMAGE. DO NOT GET IN EYES, ON SKIN OR ON CLOTHING. WEAR PROTECTIVE CLOTHING, RUBBER GLOVES AND GOGGLES OR FACE SHIELD WHEN HAND-

LING." Fascinated by the label's blunt honesty, I stared open-mouthed at the can and wondered if a few spots of wild buckwheat in the beets were worth spraying after all. Having used a small quantity of this particular chemical a couple of years ago, I had either forgotten or had never before read those startling words. So this is modern farming. With meticulous caution I carefully measured the proper amount of the black, murky herbicide into the spray coupe. It turned white upon water contact and I watched as it quickly spread throughout the tank. I sprayed with the hope that the chemical was deadly to wild buckwheat. We farmers owe a great deal of our increased productivity to synthetic chemicals and I'd hate to farm without them. But sometimes I wonder if we haven't sacrificed some sound cultural methods of weed control for a pint of this or a half-pint of that per acre.

Times have changed down on the farm. Crop rotations are less prevalent. Years ago everyone had a few cattle. Alfalfa was rotated on land, helping to control weeds. Old practices such as mowing Canada thistle just at first bloom, plowing summer fallow the third week in June, deep cultivating all summer and just prior to freeze-up on summer fallow acres to control quack grass, delaying seeding until the wild oats had grown, then cultivating them out prior to seeding wheat are examples of cultural weed control. Now farms are more specialized and less diversified. The cows got sold and the hay land and pastures got plowed. In a wheat monoculture, weeds such as wild oats, wild buckwheat, kochia and wild mustard are a more serious problem.

For good or bad, those old days are gone. We could farm without chemicals, but the toughest question would be what segment of the population are we going to let starve to death? The old cultural methods were good in their day, but yields were very depressed compared to today (and the smaller population didn't consume as much.)

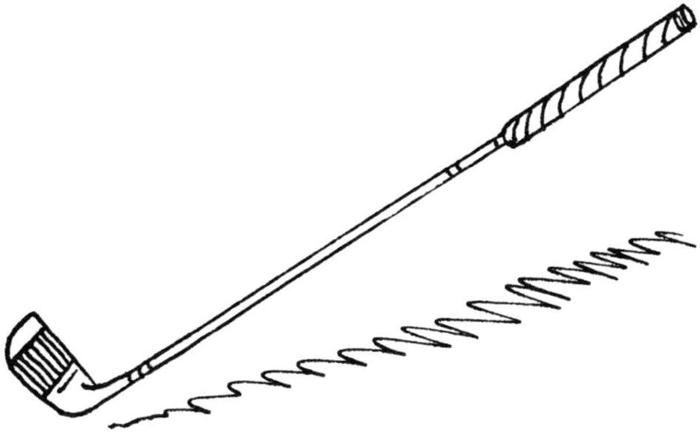

SUNDAY, JUNE 29

While Carol, Andrew and Todd are napping, I take Melissa and Alan for a little golf this afternoon. We go only five holes, but I want to start getting them exposed to the game. At each green they are given a putter and enjoy hitting the ball into the cup. Trying to teach them golf etiquette, I instruct them not to talk when other people are swinging. At one hole while I'm teeing off, Alan persists in talking.

"Shut up!" screams Melissa.

"Make me!" challenges Alan.

This prompts a knockdown, drag-out fight and requires my intervention. After things settled I instructed them on more golf etiquette: "You never get in a fist fight with your opponent."

MONDAY, JUNE 30

We spoke in quiet tones of our common occupation and the need for a rain. A lot of farmers' spirits would be raised considerably by a good soaker. I had a 9:00 appointment to certify my acres at the

ASCS office and my wait was short. In just a few minutes a welcome voice was heard.

"Dean, you're next to certify."

I sat down beside a desk piled with folders. Cleaning a space in front of her, the clerk placed a fresh set of forms.

"Okay," she said enthusiastically. "Let's start."

Minding my manners in that I only spoke when spoken to, I politely gave monosyllabic answers to her many questions. Farm number? Beet acres? Wheat acres? Barley acres? ACR acres? Landlords? Crop share or cash rent? What percent share?

Poring intently over maps, we drew field boundary lines, designating crops and acres. The information was punched into her computer and in surprisingly little time I had reported my seeded acres for 1986.

"There," she finally said. "Now all you have to do is read this and sign." She handed me a twelve-page form entitled, "Appendix to Form CCC-477, Contract to Participate in the 1986 Price Support and Production Adjustment Programs." My reading consisted of ten seconds to count the pages of small-typed definitions and regulations.

"Got it," I laughed. "Where do I sign?"

The barley is in the process of heading and is in its ragged stage. The heads are in varying degrees of protruding from the sheaths, giving an uneven and rough appearance to the field.

To allow the Poast to translocate to the roots, the directions state that a one-week waiting period is required before cultivation. However, with the wild oats turning brown and the threat of rain, I cultivated beets in the afternoon, finishing at 10:00.

TUESDAY, JULY 1

Ah, July. My favorite month. In spite of the fact that the daylight hours are waning, the earth continues to warm. The work schedule slows as both the weed spraying and beet work are nearing completion. In a few days we will have done all that we mortals can accomplish toward raising a crop. The rest is out of our control and is given over to nature. More time can be spent relaxing in the sun. Unfortunately, with this year's late spring, the number of free days will be fewer than desired.

It rained during the night and I was glad I cultivated as late as I did. The .35 inch at home and .15 at Hallock were far from what we could have used, but we shouldn't complain.

Robbie cultivated summer fallow this afternoon. The fallow has thus far been chisel plowed twice and cultivated twice. Many dollars in fuel and labor are spent to keep a field black.

WEDNESDAY, JULY 2

Sitting at my desk, piled with bills and papers, I noticed that the calendar still read May. Somehow in the process of seeding, cultivating and spraying, a month has been lost. With the "June push" over, it's time to ascertain input costs such as seed, fertilizer and chemical. Most expenses appear to be in agreement with projections penciled in last winter, except for a dramatic drop in chemical cost. Even if the crop doesn't yield as we would like, the lower expense will help to compensate.

The day was warm and sunny. The last of the weed spraying in the wheat was accomplished this afternoon.

We have three jugs of unused chemical on hand, and Alan, Todd and I drive to Hallock to return them. They love going because of the free pop machine at the dealer's. Sampling a glass of the orange soda, they move to root beer and Mello Yello, ending up at the Coca-Cola. Wearing Coke mustaches and burping carbonated soda, they ride home lying on the pickup's seat and complain of stomachaches.

THURSDAY, JULY 3

The morning was spent mowing grass and cleaning the yard. I recall a PCA loan officer once remarking how he could always tell the way a man farmed by the neatness of his yard. When my lawn is trimmed and orderly I like to think he was right. But at those times when the grass and tree rows need mowing, the garden needs weeding and the building site needs a general "picking up," I tend to feel that he didn't know what he was talking about.

A picture in the local newspaper showed what looked like the ragged remnant of a sunflower field after a severe hailstorm. Beneath it the caption read: "Hail Today, Gone Tomorrow." Utilizing scare tactics, a hail insurance company was graphically describing hail's devastation and why they were the best agency from which to buy your insurance.

Sitting by the fan at Kennedy's Hartz, Rick the proprietor panted, "I can handle ten below better than this."

The clock showed only 1:00, but the temperature and humidity were already in the 90's.

"Personally, I'm a hot weather person," I said, paying him for my groceries.

"Well, I'm sure not," he answered.

Sweating profusely, Charlie, another farmer, walked in the open door. With his cap pulled over his eyes and shirt matted with sweat, he purchased a pack of Camels. "Hot enough for you?" he asked, laughing through his teeth.

"Feels like it could rain a foot," Rick said.

"Just so nothing hard comes down with it," cautions Charlie.

"You mean that big white combine?" said Rick, taking his money.

"I bet lots of hail insurance is being written."

"Seems like it takes one storm in the county to scare everybody into buying."

Completely forgetting the farm, I buried myself in a book and lay in the sun most of the afternoon, soaking in its warmth. By late afternoon an uneasy calm settled over the prairie. The air almost dripped with moisture as a dark cloud appeared on the western horizon. Static intensified as the sky clouded and the wind switched to the northwest. The air cooled several degrees in just a few minutes. In the distance, faint thunder was rumbling. As the thunderhead approached, fingers of white clouds swirled against the prevailing wind. In the west, a dark curtain of water gradually inched in a northeasterly direction, missing us by several miles.

FRIDAY, JULY 4

The air was hot and the water was cool as Carol, the kids and I enjoyed the sun beside a crowded swimming pool. The laughter of splashing children filled the air. It's a day of relaxation and assessing the crop's progress to date.

"When's it going to rain?"

I squinted into the sun to see Roger* sitting on the ground beside me. His farmer's tan, ending at the biceps and neck, indicated many working hours in the sun.

I replied, "I thought for sure we were going to get it yesterday."

"It really doesn't matter anymore for some of the crop," he added. "Too bad. It eats on a guy."

The summer sun gradually sapped our strength, and we wearily traveled home in time to watch the Statue of Liberty's fireworks display on television. At 2:30 a.m. I was treated to a much more spectacular array of fireworks. Distant at first, the periodic low rumblings awakened me. Living on the prairie without a fully grown shelterbelt affords a panoramic view of incoming thunderstorms.

Fascinated, I watched for almost a half-hour as three or four times a minute a lightning bolt would burst the whole southwestern sky into brilliant light. One thousand one, one thousand two, one thousand three . . . One thousand ten, one thousand eleven. Rumble. Coming from the southwest was a good sign. Maybe it will track in our direction. Flash. Flash...rumble. Rumble. Flash. At a hundred million volts a crack, nature was unleashing an awesome amount of power. In spite of the fact that this was a natural way in which nitrogen is restored to the earth, it seemed like a waste of good electricity. Flash. One thousand one, one thousand two, one thousand three. Rumble. It's coming our way. Flash . . . Crash. Flash. Three miles away the town of Kennedy was suddenly illuminated before disappearing into the dark night. Crash. Flash. The yard light went out. Two miles to the northwest Gunnarsons' buildings also sat in darkness. A few minutes later the first large drops could be heard on the roof. Quickly the sound crescendoed as the eave's troughs filled to capacity. Intermittent flashes revealed what seemed to be a wall of water cascading off the Quonset. This should make a lot of people happy.

SATURDAY, JULY 5

The inch of water in the rain gauge was a welcome sight. A few hailstones also fell during the night, but ceased almost as soon as they started. General conversation of the morning breakfast crowd was upbeat, and no doubt projected yields were increased as people were hinting of new purchases. If the rain had come too late for some of the grain, the row crop will certainly be given a boost.

Walking out of the hardware store, I met Paul*. His ear-to-ear grin disclosed what was on his mind. Opening the conversation, I said, "Nice rain."

"Boy, I'll say," he replied. "It couldn't have come at a better time.

This is pretty darn good country to farm in. It always seems to come through. It may look hopeless for a while, but sooner or later it always gives you a break."

Last week this emotional yo-yo had been despondent to the point of selling the farm.

"When you put all that money in the ground," he continued, "and the gamble pays off, it's a fun game to play." Sticking his finger in my chest, he said, "If the challenge was gone, farming would lose a lot of its appeal."

Laughing an affirmative, I walked to the pickup. Momentarily I watched as he swaggered down the street, imposing his enthusiasm upon everyone.

MONDAY, JULY 7

When I embarked upon the project of originating a new building site, I told myself it would be five years before the yard was properly graded and sufficient grass was planted. It was a cold, windy day in the fall of 1977 when Carol and I measured the distances for the buildings we were planning on part of the quarter of land I bought in 1975. Trees were purchased through the Soil Conservation Service, and the little sticks that would become the windbreak were planted in the spring of 1978. Some steel bins and a metal Quonset were built at that time, and each year improvements have been made. The house was built in 1979 and the pole shed in 1983, followed by more bins in 1985. Tree planting and landscaping have occupied much time the last few years.

I am often jealous of people who live in older, established sites with fully grown trees, but there are times when living on the prairie can be rewarding. In the evening after the wind dies, it's fun to sit on the patio and listen to the meadowlarks call a quarter of a mile away. The landscape is tabletop flat. My dad often tells the kids that they'll remember growing up looking out the living room window and seeing the curvature of the earth.

I spent the morning scraping dirt from a coulee half a mile away and dumping it around the pole shed and newly built steel bins. Since 1980, much work and expense has been invested each summer hauling, leveling and packing fresh fill. A layer of rock has been added, topped by a coating of gravel. Shrubbery and ornamental trees have been scattered throughout the yard. Sidewalks have been poured, fences built, ditches cut and grass seeded. New farmyards are always being developed so this is not unusual.

As farms grow larger and more people move off the land, however, empty buildings become a common sight. The gradual destruction of them follows a sequence of declining usefulness. Trees get pushed into piles and burned. Obsolete sagging-roofed chicken houses or machinery sheds with wagon-sized doors are burned, their stone foundations buried. The house, with its rotting cedar shingles, peaked gables, broken windows and doors, follows the same fate. Finally, all that remains is a granary surrounded by a wheat field. Eventually, its rotting floor joists and leaky roof deem it unfit for grain storage, and it too succumbs to destruction. We fool ourselves if we feel our labor will not result in a similar end.

With these thoughts in mind I drove the dirt road east of Donaldson. Having gone to a machine shop near Stephen for parts, I detoured on the way home in search of an old site I hadn't visited for twenty-five years. A few faint memories led me in the general direction. As a kid I remember riding with my dad on a road not unlike the one I was presently traveling. I drove slowly, searching for clues, and the road turned south, just as I remembered. Donaldson lies five miles to the northwest, and Stephen is five miles to the southwest across the prairie.

I must be getting close. Braking the pickup, I surveyed the countryside. One hundred yards ahead, a crossing led into the field. I drove to it and stopped. This must be the place. Before me was a barley field in the process of heading. Walking from the crossing, I saw a few fist-sized rocks scattered on the ground where the driveway had once been. Picking one, I examined it carefully. An alien to the ancient lake bed, it had obviously been transported from gravel pits several miles away. The box elder and willow trees were gone. A slight rise in the topography gave notice of where the barn once was. Its concrete foundation was broken and buried under the black clay. I pictured in my mind's eye where my uncle's house, granary and machine shed had once been. The landscape had now returned to what it originally was: a flat, treeless prairie. Aspiring to the good life for himself and his family, he had worked hard planting, building and planning, but he died at an early age, his fruits relegated to insignificant mounds in a barley field.

William Least Heat Moon: "We are but tools of our dreams. Dreams that rise from the earth, to fall back again."*

The barley rustled as I waded toward the crossing. Stopping to pull a head and admiring its length, I thought, "The crop looks pretty good."

from "Blue Highway"
*©Atlantic Monthly Press

TUESDAY, JULY 8

Giving a carpet-like appearance, the barley is now in full head. Scratching my hands as I gently step through its dense foliage, the light green beards wave in the summer breeze. Depending on one's occupation, beauty constitutes a number of things. To a farmer, the sight of this barley field has to rank toward the top.

Robbie scraped dirt all day. It's a slow process, but by quitting time he had acquired quite a few yards of new fill around the buildings. After supper, I continued until dusk.

WEDNESDAY, JULY 9

In today's 80 sunny degrees, the beets are really growing. Robbie finished scraping. I spent the day removing the fertilizer tank and disassembling the grain compartment from the seed truck. A boring task that has to be done, it took until early afternoon. Changing oil in the tractors took the rest of the day. Three-tenths inch of rain fell during the evening.

THURSDAY, JULY 10

I took the day off, and did absolutely nothing except lie in the sun with a good book. Remembering the long days of spring seeding and June's beet work, and anticipating harvest, I didn't feel a bit slothful by spending the day in such fashion.

FRIDAY, JULY 11

One of the main stories this summer in any farm publication is the storage shortage. Market prices are low and most of the crop will have to be stored with government loans. Farmers, therefore, are improvising different ways to build cheap storage. With that in mind, I raised the bulkheads on the east door of the pole shed from four to six feet. Half the pole shed has been used for grain storage and half for machinery. After five hours of work and fifty dollars' worth of material, an additional 3000 bushels of storage are available.

Robbie worked summer fallow all day.

SATURDAY, JULY 12

Distending from the sheath of the flag leaf, the wheat heads are emerging. The head of the main stem appears first, and is followed in turn by heads of the tillers in the order of their origin. Walking in a later-seeded field, I break a stem from the plant. When I slice the leaf sheath with my thumbnail, a green, wet head is revealed.

A couple of fields that once seemed wild oat-free have yielded a scattering of the grassy weed. Towering above the grain, they mar the wheat's aesthetic appeal, but so few wild oats in these fields don't warrant the cost of spraying herbicide.

My family and I are filled with hamburgers, baked beans and potato salad served from local vendors as we walk among the new machinery we can't afford to buy. Congregating at the Kittson County Fair, clusters of white-shirted farmers with clean seed caps discuss wheat prices, weather and tractors. Squinting into the early evening sun, they tease each other about their machinery needs.

Prodded by restless youngsters, we hurry on our trip through the machinery. Being livestock-ignorant, the kids giggled and pointed in awe at the chickens, ducks, cattle and horses. Eventually they lost interest and heeded the beckoning lights and music of the Midway.

MONDAY, JULY 14

More rain fell during the night, another .3 inch. We've received about two inches in the past nine days. Excellent for the crop, the precipitation would have been much more helpful if it had fallen a month earlier. There's a growing concern that all the damp weather will encourage grain diseases.

Robbie and I cut and painted fence posts most of the day for a white decorative fence separating the outbuildings from the house.

TUESDAY, JULY 15

The hot, humid day is beautiful for row crops, but also great for propagating disease. In these moist greenhouse conditions the leaves remain damp. Even at midafternoon my knees get wet walking through the beets. As a preventive measure we'll spray for leaf spot disease in the beets. The disease may also dip into wheat yields by robbing some lower leaves from the grain.

After wrestling with the matter for several days, I call a local agency and take out hail insurance on the better-looking wheat. Having come this far it would be a shame to lose the crop. What a person is buying is a little peace of mind.

The sun's rays at a low angle made the wheat appear deceptively thick and lush as I drove to attend the Kennedy Farmers' Elevator annual meeting.

WEDNESDAY, JULY 16

Testing the theory that a wheat plant will produce as many tillers and heads as the climatic and fertility conditions dictate, I spent the morning walking fields. On a field where the rain had not fallen as early as desired, anywhere from one to three tillers or shoots can be found. If weather conditions permit, these secondary stalks will each produce another head of grain. In drought they will wither and become unproductive. Pulling plants for inspection, I found that many had only one decent-sized head from the main stem along with two tillers. Since they are already in the process of drying, it's doubtful whether these smaller tillers will mature. If we'd had another rain in June they would have borne seed.

On another field, where the land was better and growing conditions were more conducive to wheat, the balance of good heads to poor tillers swung in favor of the former. The majority of plants seem to have three heads with only one non-producing tiller. One plant, the exception, had six heads. Too bad the whole crop can't be like that.

Having devoured the lower leaves, the wheat diseases are gradually creeping up the plant.

While I was sitting on the pickup endgate with my wheat plants spread before me, Donald* drove up from the east.

"Well, look here," he said, rolling down the window. "This reminds me of when Adrian Olson* used to count the kernels in the wheat heads and figure he was going to get such a terrific crop that he'd run to town and buy a new tractor."

"It's not what you think," I laughed. Pointing to the yellow, disease-infected leaves, I asked, "Have you sprayed any Dithane?"

"No. Maybe we should have," he replied. "But doggone it, wheat's just too cheap to keep throwing money into. If it was four or five dollars a bushel, it'd pay to spray. At $2.30, or whatever, we'll have to let the disease take a few bushels."

"I agree. The economics just aren't there."

After he departed I drove to a beet field and pulled an average-sized root for measurement. At two and a quarter inches in diameter at the crown it wasn't too bad for this time of year.

THURSDAY, JULY 17 to SUNDAY, JULY 20

Carol and I reserved a cabin for our family at Bluewater Bible Camp near Grand Rapids, Minnesota, one of the most relaxing places in the world. For four days we soaked up the sun. The Camp is nestled beside a beautiful lake and surrounded by lovely trees. The mornings are spent at church services, while the afternoons are for recreation such as boating, swimming, water skiing, sunbathing and golf. In the evenings church services are again held. The meetings are inspirational, and we get a chance to make new friends. The kids love the water, and Carol likes the fact that there are no dishes to wash and no meals to prepare. Meals are served cafeteria-style with more food than we can possibly eat.

Sitting beside the lake after a swim, I was joined by another farmer.

"Isn't this great!" he exclaimed.

I had to agree.

"You know," he added, "when I'm here, I completely forget about the farm. When I see thunderheads approaching in the sky I realize that we're so far away from home that there might be a different weather system back there. I don't worry about what is happening at home at all."

The weather was terrific, and Carol, the kids and I all had a good time.

MONDAY, JULY 21

For all farmers coming home, the first order of business is to drive around the section inspecting fields. My short absence seemed longer, and I half-expected to see some dramatic change in the crop. The only change, however, was in me. Feeling refreshed from the break, I'm anxious to resume some needed repair work before harvest.

Robbie cultivated summer fallow while I mowed roadsides.

TUESDAY, JULY 22

It's hard to believe that July is two-thirds gone. While Robbie finished some summer fallow work, I added more posts to the decorative fence. It was one of those jobs that if you had known how much work it would involve, it would never have been started. By midafternoon the project was completed.

In today's 86-degree temperature, the barley is rapidly ripening. Its green color is fading, and shades of tan and yellow are becoming more prevalent. We should be swathing about August 10th. When that happens, fall officially will be here and the lazy midsummer days will be over.

The air conditioner dripped water as Mike Gunnarson and I stepped into Gary Hultgren's shop. An expert mechanic, he was bent over an engine with sweat pouring from his brow.

"I couldn't stand to be out here without that," he said apologetically, pointing to the humming compressor.

As township board members, Gary, Mike and I lazily drove the township roads, inspecting them for needed gravel or repair work. Everywhere the ripening grain indicated the upcoming harvest. Occasionally inspecting culverts and reviewing a portion of Tegner township's history, we swapped stories of ancient arguments between neighbors over water drainage.

WEDNESDAY, JULY 23

Another .25 inch of rain fell during the night. After the clouds cleared, the day became hot and humid. These are ideal growing conditions and the beets are thriving.

THURSDAY, JULY 24

In an effort to obtain as much storage space as possible, Robbie and I began the task of transferring grain. The contents of smaller bins will be emptied and piled in the pole shed, where machinery is usually parked. I hate to see the grain drills and beet planter sit outside over the winter, but this year it can't be helped. The beet thinner and combine can be squeezed in the front of the Quonset.

The various stages of development of different crops create a colorful landscape. While the barley is just ripening, the wheat and beet fields remain green. The flax is in full bloom and appears as a sea of purple in the mornings.

The wheat plants' uppermost internodes have elongated to maximum height. After the flowering process the tiny reproductive parts litter the ground.

FRIDAY, JULY 25

The breakfast conversation at the local restaurant was more animated than usual. Combines and politics were the main topics, when a farmer wearing a dirty Avenge herbicide cap walked in the door.

"What's for breakfast?" he demanded.

"Same old junk," came the reply from the kitchen.

"Gimme a plate of the same old junk," he said as he sat down. A heavyset jovial man grabbed Avenge in the buttocks as he walked by.

"How ya doin', Sport?" he shouted.

"Watch it," Avenge warned.

Bringing cutlery wrapped in napkins, the waitress banged them against the tabletop.

"That's to get the dirt off," she explained.

Laughter among the crowd. The sizzle of bacon in the kitchen could be heard, and as always, "Good Morning America" blared from the small black-and-white TV. An egg over-easy was placed in front of me. I've discovered that more news is spread, yields announced, gossip exchanged and political opinions dogmatized over two eggs and toast than by any local radio or newspaper. Suddenly the conversation is interrupted by a slender man wearing a crooked Far-Go herbicide cap.

"Well, it's that time again. You gonna buy me breakfast, young guy?" Shaking the dice box in his hands, he looks at a mustachioed farmer wearing an American Crystal cap.

"We'll see about that," Crystal replies. "Roll 'em."

Conversations cease as all eyes are glued to the five ivory dice rolling from the black cylindrical container. Rattle.

"Doggy, I'm hot today!" Far-Go announces.

Leaving three dice, he picks up the other two and rolls again. Rattle. A six and a two are face up.

"Watch this." Leaving the six and depositing the last die in the box, he shakes again. Slam. The box sits face down on the table.

As he slowly lifts one side to peer beneath it, his countenance lights as he shouts, "There's another!" Imitating a magician pulling a rabbit from a hat, he exposes the second six for all to see. Laughter and headshaking from the gallery. "Don't get too cocky," cautions

Crystal. Grabbing the dice and shaking, he hypnotizes the audience.

For the next couple of minutes Far-Go and Crystal have center stage. All is silent except for the rattle of dice accompanied by the grunts, groans and fragmented sentences of the participants. Crystal rolls well, but the gods of the dice box are with Far-Go this morning, as he gets a free breakfast.

I brought home a considerable amount of two-by-fours and plywood, and started cutting and nailing the first bulkhead in the pole shed. It took most of the afternoon, but I was confident that once a pattern was established the rest would go quicker.

The humid 85-degree temperature is good thunderstorm weather. At 5:00 clouds appeared in the west. Switching to the northwest, the wind cooled the air as I watched the thunderhead approach. The light patter of rain was heard on the pole shed's metal roof, and then suddenly the heavens seemed to open. Laying aside my hammer, I stepped to the door and realized the tin roof had greatly magnified the severity of the rain. The total amount was a half-inch.

SATURDAY, JULY 26

A perfect day. Golf in the morning and my annual aerial view of the country in the afternoon.

Carol, the three older kids and I are treated to an airplane ride by my brother-in-law, Rich Anderson. Squealing with excitement, the kids gawk at the countryside far below. We fly to the Red River, twelve miles west, pointing out familiar building sites. Circling, a few minutes later we are twenty-five miles east, looking for deer and moose in the swamps of eastern Kittson County. A lone water skier leaves a white wake on Lake Bronson. From a thousand feet the crop never looks as good as from the pickup, because every wheel track and pothole are seen. Black spots twelve rows wide remind me of the times the beet cultivator slipped off the row. Beautiful from the air, the land is a patchwork of greens, yellows and black. The sugar beets are a lighter shade of green. After an hour's flight, and with my stomach feeling a bit queasy, I was glad to have my feet on solid ground.

MONDAY, JULY 28 TO WEDNESDAY, JULY 30

Hoping to cram several thousand bushels of grain into the pole shed, Robbie and I spend three days building bulkheads. Having our separate jobs, we work mostly in silence. I build the framework while he cuts and nails the plywood. The sound of hammers and saws is all that we hear. Quitting at 6:00 p.m. each day, I painfully walk to the house with a new appreciation of how hard carpenter's work is. Our best day was Tuesday when we built ten of the eight-foot sections. Late Wednesday afternoon all are completed.

THURSDAY, JULY 31

The wheat is in the soft dough stage. Squeezing the kernels reveals soft, white starchy material. The bottom leaves are completely covered with disease. Even the flag leaf has a few lesions.

Spraying for leaf spot disease in the sugar beets is a contagious thing: Once a few growers start all the others panic, which results in everyone's spraying. At $7.50 an acre for one application, the cost isn't too exorbitant.

FRIDAY, AUGUST 1 TO SUNDAY, AUGUST 3

We rented a motor home and drove to a resort near International Falls for a three-day weekend. Owned by Carol's aunt and uncle, the resort is set beside a beautiful lake. Rain hampered our outdoor activities, and we found ourselves confined quite a bit of the time in the motor home, but the change of scenery was refreshing.

MONDAY, AUGUST 4

Although it was busier than anticipated, the month of July was enjoyable. Now that harvest is near, Robbie begins arriving for work at 7:00. My self-motivating list of "Jobs To Do" once more graces the shop wall. Not accountable to human employers, farmers are subjugated to intangibles. Self-esteem, peer pressure and nature demand that the work gets done.

WEDNESDAY, AUGUST 6 TO THURSDAY, AUGUST 7

These days are spent working on projects in-and-around the yard. Trucks need to be serviced, bins swept and the dryer set in position. One of the biggest jobs is repairing the swather. Last year's swathing in the snow and on frozen ground gave it a terrible beating. Many sickle and guard sections have to be replaced. The list is slowly diminishing.

"Dad, come quick! Mom hurt her back!" Melissa shouted as she ran into the Quonset.

Arriving at the house, I found Carol stooped over, unable to straighten her back. She had been lifting Andrew and turning to the side when something slipped in her lower back. We called Hallock's chiropractor and made an emergency visit, but her back would require several more treatments before she felt better. Carol was in much pain for the next few days, and I became more appreciative of all the work she does around the house and yard. With much pre-harvest work to be done, I found I couldn't help as much as I wanted to with the household chores. Various people stopped over to help, and we were very grateful to them.

FRIDAY, AUGUST 8

The day has arrived. I drove the self-propelled swather the two-mile trip to Section 23 and began cutting barley, but after one round I thought the crop still seemed a bit green. Parking at the edge of the field, I loosened the canvases before driving home. Next week will be better.

Robbie and I hauled 2,500 bushels of wheat to Kennedy in the afternoon.

MONDAY, AUGUST 11

After a drizzly weekend, the sun shone this morning. Trying to make a living in seven months, people everywhere are busy. The large steel bin at Kennedy Farmers' Elevator is under construction and growing taller daily. Fertilizer companies are repairing anhydrous wagons and floaters. Implement companies and hardware stores are doing a brisk business. Working after normal quitting time, mechanics at Massey Ferguson repair a 760 while the proprietor of the local radiator shop sweats over a Versatile tractor.

A few farmers have begun combining barley.

TUESDAY, AUGUST 12

Using the formula given me by a crop insurance adjuster, I walk fields, attempting to estimate yields. If the pole shed is used, we should have enough storage. My own crop's worst critic, I habitually underestimate the yield. That way, if it produces a few bushels more, it seems like a bonus.

In the afternoon I cut the barley. Some places were a little green, but with the warm weather it will cure in the swath.

Robbie began seeding our set-aside, or ACR acres, in oats. A wise stipulation of the farm program is that this acreage must be seeded to a cover crop, thus protecting the land from wind erosion next winter and spring.

THURSDAY, AUGUST 14

Robbie finished seeding oats today. I checked barley swaths, and the straw felt a little tough. Good weather is in the forecast, so I decided to try combining tomorrow. If we were to swath wheat, a couple of fields could be cut. Wanting to straight combine, I decided to wait.

For economic reasons, quite a few farmers have held off spraying a lot of herbicide, and they are now lamenting their dirty fields.

Because last year's wet fall prohibited combining, some fields are being worked for the first time this week. The heavy layer of straw has kept the ground at saturation levels all summer.

FRIDAY, AUGUST 15

Better than any man-made dryer, the 85-degree temperature and strong south wind brought the barley moisture to 12.1 per cent. Totally disintegrating as it moved through the combine, the brittle straw is carried aloft as it spews from the chopper. It's fun to be threshing on such a beautiful combining day. After several stops for setting the sieves, the 7720 steadily devours the swaths.

A neighbor stops his pickup on the adjacent road to watch us. Loading on the go, Robbie drives the white twin screw under the outstretched unloading auger. I pull the unloading lever and barley immediately begins pouring into the truck box. Unaccustomed to this particular truck, Robbie is in too high a gear and kills the engine.

Seeing what was happening, but not reacting quickly enough, I dump a pile of barley on the truck's hood and cab before disengaging the auger. The neighbor politely drives away from this embarrassing situation.

The barley is yielding quite well. Only a short afternoon is required to combine the thirty-five acres.

SATURDAY, AUGUST 16

I tinkered in the shop while Robbie tandem disced the barley ground. Finishing that, he chisel-plowed it with the twisted shovels. Quitting my work at 8:00, I'm glad to sit down. It will be nice to start combining steadily. At least then I'll be able to get off my feet.

Purchasing parts at John Deere, I ran into Wallace*.

"With these barley prices, the maltsters are making a fortune," he said. "If they're purchasing the barley for a dollar less than last year, the price of beer should also drop."

A thinking man, he calculated how many pints of beer are made from a bushel of barley, and offered a figure as to how much a bottle of beer should drop in price. His rationale seemed sound.

MONDAY, AUGUST 18

Bob, Robbie and I built the doorway bulkheads for the pole shed today. Many times I've wondered if these things are worth the work. Tomorrow we'll begin filling.

TUESDAY, AUGUST 19

Robbie and I finally started to fill the pole shed. Transferring wheat from a steel bin, we trucked it to a waiting auger aimed at a far corner. As each load is dumped, the cone-shaped pile of grain gradually enlarges. Within the pole shed the auger's noise is deafening. Dust billows from the open doors. Trips into the building to check the pile's height, or to raise the truck box are hurried. When the wheat reaches the top of the bulkhead the auger is swung to the side, filling another portion. By midafternoon a 4,000 bushel bin was emptied.

Champing at the bit, I've been driving past a thirty-acre piece on the Home quarter for a few days, longing to be combining. It appears ready from the road, but underneath the ripe heads are the green, immature plants which germinated later in the spring. The question I'm mulling is whether or not the percentage of green kernels will be great enough to cause spoilage in storage. Only one way to find out. I made the half-mile trip with the combine to the field crossing. I engaged the separator and the combine came to life with a mild vibration. Then I pulled the feeder switch and the twenty-two foot sickle and reel went into action. Slowly at first, the 7720 cut a swath in the standing grain. Increasing ground speed, I aimed for the center of the field. Grain rattled into the hopper after it made its way through the bowels of the combine. The first 1986 wheat was being harvested. Swinging into a broad turn, I cut a loop in the standing grain. Ending where I had started, I ran the separator at full rpm for several moments, emptying the combine of straw and grain. The moisture tester read 13.6%, dry enough to store. The abundance of immature kernels, however, gave the wheat a green hue. Squeezing them yielded a soft doughy starch. Uncomfortable with binning such wheat and disappointed that straight combining could not be done, I drove home to get the swather. The grain will have to cure in the swath before I dare thresh. Three hours later the field was covered by parallel twenty-foot windrows.

WEDNESDAY, AUGUST 20

Another beautiful, sunny day in the 80's. The straight header was removed from the combine and the swath header attached. After dinner another sample was threshed. Wishing to compare my moisture tester with that of the elevator's I ran a coffee can full of wheat to town.

"Yuk. It doesn't look very good."

Not appreciating his lack of tact, I replied dryly, "Tell me something I don't already know, like moisture content."

Embarrassed at his undiplomatic statement, the elevator manager replied, "It's the same everywhere. Every sample we've got in looks poor. The disease has really dipped into the quality."

He was right. Even an untrained eye could detect the shriveled kernels. Sprinkled throughout the sample, a few whitish-colored seeds gave notice of the wheat scab disease so prevalent this year. The contents of my can were cleaned, weighed, and poured into the moisture tester. After turning a dial and consulting his chart, the manager returned the verdict.

"12.5% with lots of green."

It had dropped a point since yesterday.

"Does a man dare bin it?" I asked.

"People are," he carefully answered, "but I'd keep an eye on it."

Entering the room, an employee sniffed disdainfully at my wheat and said, "Looks like what Ralph's* hired man brought in." Shaking his head and reaching for a cigarette, he continued, "He said they had straw for a fifty-bushel crop but it was only yielding thirty. I think there's going to be a lot of people disappointed in their crop."

Unlike last year, it was fun combining from end to end without getting stuck, and it took but a short afternoon to pick up the waiting swaths.

THURSDAY, AUGUST 21

The morning dew was wet on my boots. The situation was the same as yesterday but on a different field. Before me sat a quarter section of seemingly ripe wheat. Spreading the straw revealed the problem at hand: green heads. What to do? Wait for all the green heads to mature? What percentage of the yield do they represent? Probably quite small. Are they worth waiting for? It was the third week in August, and combining had barely begun. Can we afford to keep waiting? Should the quarter be swathed or straight combined?

A closer inspection divulged the fact that the wheat plants with remaining green heads were completely without leaves. Will those kernels fill any more? Swathing would accelerate the ripening process but would involve the risk of losing quality should the weather turn rainy. It's a risk that will have to be taken.

Arriving at the farm at 8:00, Robbie was awaiting instructions as I drove into the yard. Fueling the self-propelled swather, he commenced the three-mile trip to Hilmer's to start cutting.

Bob is a different story. He's back from his custom combining job and I never know when he's going to show up for work. Today it was 8:25, and as usual another feeble excuse explained his tardiness. I expounded that I would not tolerate his habit of coming to work late. He seemed repentant, and under the threat of being fired promised it wouldn't happen again. He spent the day discing and chisel plowing the newly combined wheat stubble.

Due to the disease infestation we are planning to moldboard plow everything this year except those pieces where sugar beets will be planted in 1987. Moldboard plowing uses the everyday plow we all are familiar with. A slab of ground gets cut and turned upside down by a moldboard, different from a chisel plow, which is a heavy-duty implement used to till the ground but doesn't completely bury the trash. With a moldboard plow, disease organisms and weed seeds are buried under six inches of dirt where they will be less likely to affect next year's crop. The drawback is that the land has less straw cover for wind erosion over the winter.

Late in the morning I saw the swather coming from the north. As it loomed closer, I wondered what had broken. If it wasn't serious it could have been fixed in the field. Driving it home was a bad omen. Knowing what I was thinking, Robbie swung the left side of the platform in front of the open Quonset door. Immediately I could see the problem: the frame had broken and would require extensive welding. By midafternoon, after much straightening and welding, the repair was accomplished.

FRIDAY, AUGUST 22

The soft patter of rain on the roof awakened me in the morning. Calling Bob and giving him the day off, I waited for Robbie. Replacing a bearing on the swather required removing the three main drive belts, plus disassembling much of the drive train. Heavy clouds dripped just enough rain to get us wet. By noon, the swather was ready to work. Sending Robbie home, I cleaned the shop. Warm, dry weather is needed to harvest this crop.

My "Help Wanted" ad in the local newspaper has been getting some response. We need extra help for beet harvest. We've been fortunate in previous years to secure good drivers through the want ads. Experience is a secondary prerequisite to dependability. If a person comes to work, I can teach him.

Three seedy-looking characters were in the car as it slowly pulled into the yard. As the car stopped in front of the Quonset the driver got out and asked rather sleepily, "Are you Dean Carlson?"

"Yeah," I ventured warily.

"I heard you were looking for beet truck drivers."

"Sure am."

"Well, I'm looking for a job."

All the while he was taking inventory of the shop. I had the feeling I should be locking the door after this. His sidekicks sat smoking in the car.

"What are you paying?" he demanded.

"Six dollars, if you've got experience. Have you driven truck for anyone else?"

He walked back to the car, put both arms on the top and stared off in the distance. "I think I worked for Severin Spilde."

"You THINK you worked for Severin Spilde?"

"Yeah, man, I drove potato truck for him."

"Jobs pretty hard to come by?" I asked.

"My old man says he could get me a job at the beet plant, but . . . there's some dirty jobs over there," he yawned.

By now I had heard enough and dismissed him by saying that I had to save the job for a driver from last year. This seemed to be just the thing he was waiting to hear as he quickly stepped into the car and drove off.

SATURDAY, AUGUST 23

The morning was foggy, with a heavy dew. It was almost noon before Robbie started to swath. After finishing Hilmer's he moved to Hallock and cut Clara's 60. Finishing that, he did twenty acres on Dad's home quarter. The wheat by Hallock is terribly poor and will be a crop insurance adjustment. Containing so many green heads and a general mess of wild buckwheat and pigeon grass, the swaths will have to lie some time before we can combine.

Perhaps it wasn't the most polite thing for me to do, but thinking it was just another salesman, I hid in the kitchen with my morning coffee while Carol answered the door.

"I've got something to show you," he said.

"Say," she said. "I really like it. Dean, come see this."

I walked around the corner to see a man standing in the doorway holding a large picture of the farmstead. It was, as Carol stated, a very nice picture of the buildings.

"You're right. That IS nice. How much is it?"

"Ninety dollars with the frame."

"Too bad we bought one two years ago," Carol replied, pointing to the picture hanging on the wall.

"I see you've made quite a few improvements since then," he answered.

We had put up more bins, seeded more grass and had done some cosmetic things around the buildings. His picture had also been taken at a better angle.

"Too bad it's so expensive," I said. "Sorry, but we'll have to pass on this one."

"Why do you come around so often?" Carol asked.

"It was another outfit that took pictures here a couple years ago," he replied. "But we come and take pictures because this is probably one of the more affluent areas in the country, here in the Red River Valley." He smiled sheepishly and added, "People are more apt to buy them here."

Whether it was a good sales pitch or not I don't know. "I'm flattered," I replied, "but no thanks anyway. It's still ninety dollars."

As he drove away I was reminded of an old college professor who made the prophetic statement that the Red River Valley, due to its rich soil, would one day be used strictly for specialty crops. Whether you agree with the professor's prediction or not, you have to concede that this valley has been extremely good to a lot of people.

MONDAY, AUGUST 25

The weatherman reported a sixty-percent chance of rain but when the wind switched to the northwest, I felt the threat was over and we continued to swath. Robbie refueled the swather while I greased.

"I'm down to a couple sickle sections," he said.

"Been breaking a few?" I asked.

"Yeah, quite a few. There's so much green junk, and I've gotta run the platform so low."

"I'll get you a few more."

It was while I was in Hallock getting sickle sections that I met Brian* perched on a stool at the Versatile dealership.

"How's the crop?" he asked.

"Some is good. Some isn't worth a darn," I answered.

"Do you carry crop insurance?"

"Yes."

"You know, if it wasn't for the fact that a man has to get the ground worked for next year, you'd be better off leaving it." Pointing at me, he continued, "Figure it out. If you combine the wheat and sell it for $2.40 or leave it in the field and collect $3.30 from insurance, that means for every bushel you combine, you lose ninety cents."

"I can't argue with you. It doesn't make sense. But like you say, there's next year."

On the first sample run through the combine, the moisture tester read 15.6%. Deciding to take it wet or dry, I threshed one-and-a-half loads on Hilmer's before hearing a thumping sound under the cab. Something wasn't right. Where's it coming from? With the engine idling and the separator engaged, I searched under panels, listened to pulleys and examined belts. Finally the problem was isolated by squinting into the dark, dusty cavern under the engine compartment. A worn countershaft bearing was the culprit. But how do you get at the darn thing? Let's see . . . this bracket comes off. These belts. This shroud. How do these pulleys come off? Confused, I called John Deere for help. It was just as well I quit. The moisture had risen to 19.5%, quite a bit wetter than I wanted.

TUESDAY, AUGUST 26

The forecast is for cool weather the rest of the week. It appears as if we'll be drying lots of grain. It must be fall. This morning I donned a T-shirt, shirt, sweatshirt, coveralls and jacket—five layers. I guess I'm just not accustomed to the cold.

The day is cloudy and we start combining at 2:30. Several warm, sunny days will be required to ripen the wheat. Scattered throughout the countryside are newly-plowed fields. Seeing them makes me realize that the last process has begun.

While combining, I noticed Robbie was slow in getting back with the empty truck. He must have run into a problem. Considering his mechanical ability, it must be a serious one. Finally the last truck was filled, the combine's hopper was heaped and I started for home. Halfway there I met Robbie. He reported the dryer wouldn't ignite and the wet auger's bottom bearing was getting so hot it was charring the grain. I took the truck and Robbie went to the combine.

All the wiring on the dryer looked intact: propane was shooting into the combustion chamber and the spark plug appeared good. All of a sudden it ignited and worked fine. Although it was a relief for the dryer to function properly, I don't know what caused the problem and won't know until the next time it decides not to work. I planned to work on the auger's bearing in the morning.

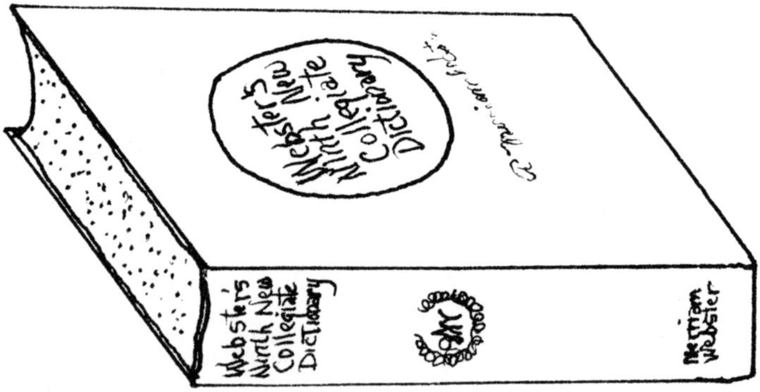

WEDNESDAY, AUGUST 27

According to Webster: "STRESS: A physical, chemical or emotional factor that causes bodily or mental tension."

If the most popular subject of farm periodicals is how tough things are down on the farm, the second is stress. Farm psychologists are having a field day defining, describing and finding ways to avoid stress. On paper it seems simple. All you do is RECOGNIZE the symptoms and REMOVE yourself from the source or situation.

The day dawned bright and sunny, although a bit cool. Thinking it could go to combine, Robbie and I worked feverishly on the wet auger. After pulling off the bearing I drove the thirteen miles to Hallock for a new one. Arriving home, I found I had bought the wrong size and turned right around for a second trip. Then of course the dryer would not ignite so I spent the whole morning trying to fix that as well.

Remembering the simple two "R" plan (RECOGNIZE and REMOVE) of combating stress, I clinically approached the problem at hand. Step #1, RECOGNIZE. I certainly recognized the headache and tightening of the shoulder muscles as stress-related. So far so good. Step #2, REMOVE. This proved more difficult. I could remove myself from this cursed auger and dryer by going to town and trading for new ones. This sounded appealing but my checkbook lacked the $20,000 required for such a move.

Another option was backing both of them into the pond. In the mental wrestling match, this irrational temptation almost pinned sound logic. There was only one thing to do. I set my jaw and continued the repair work. Sometimes we have trouble putting things in perspective. Next week—no—tomorrow morning, things will look better and the misfortunes of today will be forgotten.

Stress affects all of us. Sometimes it isn't the ugly demon so often imagined and written about, as it can increase productivity. The problem is in proper management. Whoever develops a process in which stress is bottled and drunk in proper amounts, thus releasing the desired adrenaline, will not have to work another day in his life.

Finally getting the auger fixed and the dryer working properly, we started combining at 3:00. Three hours later some green swaths were encountered and we were forced to quit.

THURSDAY, AUGUST 28

Where has summer gone? The nights are cool and the dew is heavy on the grass. This morning there was actually frost on the roof, unusual for this time of year. The weather forecast is predicting dry weather for the rest of the week, with a chance of rain on Sunday. I hope the rain will miss us. Today is just what we need: sunny. A large portion of the morning was spent setting the plow for Bob, which involved frequent adjusting and readjusting until the proper depth was finally found. The ground is dry and hard, producing large lumps as it's moldboarded. Later in the afternoon a few combines begin to blow dust, but I console myself that the good forecast tomorrow will bring a better day. Maybe Mother Nature can do some of the drying for us.

The ripe heads were rustling in the breeze as I stepped through the standing wheat. This particular eighty-acre piece on Section 11 had germinated uniformly, and I was hoping it could be straight combined. Picking a head, I threshed it in my palm. Transferring the crushed head from hand to hand while lightly blowing, I separated the chaff from the wheat. When I bit a kernel it snapped between my teeth.

"Darn it anyway," I breathed to myself. "Why aren't I over here straight combining instead of horsing around with those swaths?"

My watch showed 7:00 p.m.. Too late to start today. Tomorrow it should go well.

FRIDAY, AUGUST 29

Both Bob and Robbie came to work at 8:00. We hurriedly serviced the combine, switched headers and were straight combining on Section 11 by 10:30. As expected, the first couple of rounds were a little green and will have to be dried. Further into the field, however, the moisture dropped to where the wheat could be binned directly. Not having to run the dryer is a good feeling. The day is sunny and in the middle 70's.

Leaving half-moons of standing wheat on each corner, the 7720 continues to encircle the field. Gobbling the ripe wheat at 4.5 mph, we are threshing what will probably be our only fifty-bushel wheat of the year. Whenever possible, the combine is unloaded on the go. Driving underneath the extended unloading auger, Bob carefully keeps the truck in line with the moving combine. Pulling the unloading auger lever, I watch as the hopper quickly empties. When the unloading is complete, the auger is swung back and Bob steers away from the combine in a broad circle. Driving through the dust, he positions himself for the next round. With each round becoming a little shorter, the hopper fills more slowly until finally two rounds can be made before dumping. Supper is brought to the field at 5:30. At 6:45, the last strip in the field's center is cut. After cutting the half-moon corners, we go home, switch the headers and move to Hilmer's quarter to pick up the waiting swaths.

As annual plants die and perennials prepare for winter, the once-lush, green countryside is gradually transformed into colors of death and dormancy. My lawn is now hard and brown. Air-borne Canada thistle seeds are being carried to next year's unsuspecting seedbeds. In the ditches, the curled dock is reddish brown and the goldenrod's yellow is a tell-tale sign of fall. A few cottonwood trees are sprinkled with tints of yellow. Vehicles leave long dust trails as they speed along graveled country roads. With the exception of the sugar beet and sunflower fields, the huge rectangles and squares of the once-growing grain have been metamorphosed into tracts of brittle straw. Finally with the fall tillage they return to their original blacks and grays.

As evening arrives the breeze dies, and the dust trails no longer blow away from the combine but are suspended where they are spewed. Farmers knowingly refer to this phenomenon as a sign of dew. Unless the grain is to be dried, combining will have to cease early. The hot days so conducive to threshing are over. To the west, Gunnarson's two green 8820's are at work. Two miles east, Alden Hagen's 860 Massey Ferguson can be seen, and a mile beyond him a red blur with a dust trail depicts Lamar Johnson's Massey. Southward two miles, Kennard and Evans Anderson's red Massey and yellow New Holland are chewing straw in the gathering dusk. As darkness settles over the land, one by one their combine lights are turned on. Eventually they disappear. Midnight arrives and I wearily pull the fuel shutoff. In the ditch, the cottonwood trees rustle slightly as I grope my way in the starlight to the silently waiting truck.

At home, today's mail includes a letter from American Crystal. We start opening beet fields on September 24. Perhaps the biggest change in my farming career came in 1984 when I purchased 100 acres of beet stock in American Crystal Sugar. Summers and falls are now more hectic, but beets have been a fascinating crop to raise, and I have added beet acreage whenever I can. The time clock begins. We have three-and-a-half weeks to complete grain harvest and fall tillage.

SATURDAY, AUGUST 30

The 7720's fittings are greased. Belts and chains are checked for tightness and wear. Compressed air is blown through the radiator, freeing the cores of yesterday's dust. Windows are washed and fuel is added. In anticipation of wear, shafts and pulleys are shaken and tested. Sieves and strawwalkers are cleaned. The engine air filter and air conditioning filter are blown clean. On ignition, black smoke pours from the exhaust as the combine's engine momentarily races before settling at a soft idle. My watch reads 10:00. With any luck, quite a few more acres will be done before day's end. Labor day weekend is upon us and the city people are taking their last long holiday of the summer. The morning talk shows are discussing safe driving and how to winterize your lake cabin. I feel as though I'm getting away with something as the wheat is dry enough to bin and the dryer can sit quietly. Hilmer's quarter is finished at 1:30.

Next on the agenda are sixteen acres of wheat surrounding beets on Dad's 40, ten miles away. The huge swaths are soon being eaten. There is so much straw on this piece I fully expected the wheat to yield fifty to sixty bushels in spite of the disease. When the hopper didn't fill as fast as it should have, I suspected something wrong. Stopping the combine, I carefully inspected all trap doors and was crawling on the ground searching for seed when Robbie drove up with the truck.

"Throwing over?"

"I wish I was. There's just not any wheat here," I complained.

Finding a suitable spot between the combine's fresh tracks, I carefully brushed the straw and gently blew away the chaff. A few insignificant wheat kernels appeared.

"No wheat," I observed.

Robbie grunted an agreement.

"Why don't you take it for awhile so I can check behind?" I asked.

Robbie ran the combine and I walked behind the chopper with a grain shovel, catching straw from the walker. In its contents no wheat was found. That's not the problem. Running to catch up, I caught a shovelful of chaff blowing over the sieves. Once again the shovel yielded nothing but chopped straw and hulls. The combine was doing a good job. There just wasn't the yield I was expecting. As it turned out, the field would produce a disappointing forty bushels per acre.

After finishing Dad's 40 we moved across the road and commenced combining his home quarter. The ground is dry, and large crevices have developed in the black clay, making plowing difficult. The coulee is dry, and what was a little pond on one of its meanders is now cracked and empty. We could use a rain to make plowing easier but we don't dare breathe a word of it for fear it will start and never stop.

Toward dusk an airplane soars from the Hallock airport. As its circling pattern carries the pilot directly overhead, he tilts the wings in a pilot's salute. Flying to the northeast, he eventually disappears over the distant trees. One at a time the mercury vapor yard lights of neighboring farms start to flicker, and after a few seconds of seeming indecision, come to life. At first their radiance appears faint but within minutes they glow brighter until the surrounding buildings are illuminated. A mile west, the drive-in theater attracts a string of cars. A few latecomers racing from the east switch on their headlights. I do the same with the combine. The instrument panel shines a pale green in the darkening cab.

Blackness gradually envelopes the countryside. Hidden in the moonless night are the familiar trees, buildings and power poles. My world is restricted to the few feet in front of the combine's lights. Direction is maintained by the known positions of yard lights. To the north and east, the lights of distant combines moving slowly across the black background can be seen. The glow of their support trucks' clearance lights is a stationary orange dot. Occasionally the truck's headlights appear and merge with the combine's. Another hopper is being dumped. A few minutes later they separate. The truck's lights disappear and its driver is sitting alone in darkness. I wonder if they're speculating about me as well. Time goes faster when there's company.

"Dean, do you have a copy?" Sitting at the end of the field, Robbie has arrived after dumping another load.

Reaching for the CB, I reply, "Go ahead, Rob."

"I've filled that one bin. Do you want me to move the auger?"

I look at my watch. 10:00.

"No. Why don't you take the pickup home. I'll fill the two trucks and quit. See you Monday morning."

About 10:30, headlights appear in the south. As they drive by, I recognize Kenneth Pantzer's trucks, followed by a pickup. They must be calling it a day. I watch as they drive another mile toward the inviting light of an open pole shed. As if waiting for Kenneth to quit, the other neighbors suddenly do likewise. Living room and kitchen lights start to extinguish. I find myself checking my watch more often. Loneliness creeps over me. To ease the boredom I carry on a semi-intelligent conversation with myself. The local radio station's transmission has become a static blur. Finding a Winnipeg station with a strong signal, I listen for awhile before turning it off. The sounds of the combine are my only company.

"Maybe one more hopper," I say to myself, looking once more at my watch.

Using the yard lights as landmarks, I attempt to judge my coordinates and guess the distance to the field's end. Suddenly the swath's end appears on the headlights' periphery. Lifting the header, I swing the machine in a neat 180-degree turn. Briefly flashing on the neighboring woods, the lights pick up two sets of watching eyes gleaming in the darkness. Abruptly the next swath bursts into view. Quickly I slow the ground speed and drop the header. The hungry pick-up fingers race the waiting swath into the throat of the feeder housing. Devouring the straw, the 7720's engine groans ever so slightly as the cylinder fills, before it stabilizes at a constant drone. The digital engine rpm gauge reads 2240. My watch reads 11:20. Fifteen more minutes before another hopper is filled.

"Maybe I oughta quit with one more."

Fatigued, I turn the radio on. An all-night talk show describes how to trim your geraniums. I turn it off. The engine's constant muffled roar and the muted whine of hydraulic pumps become more apparent. Inspecting the straw chopper's spread pattern requires a quick opening of the cab's door. The high pitched scream of a dozen pulleys contrasts with the lower-toned steel-against-steel meshing of chains and sprockets. Shafts, augers and fans wail their accompaniment. The buzz of the straw chopper adds its share of decibels to the deafening noise. I slam the door, my appreciation for "quiet cabs" rekindled. The hopper gets dumped. Another. Then another. Finally the second truck is filled. I position the combine at an angle perpendicular to the road and cool the engine before shutting it off. Sitting in the tranquility of the cab, I reflect on another "good" day. One hundred acres combined. After a day of noise, the night's quietness soothes my ears. Finding my lunch kit, I walk to the truck.

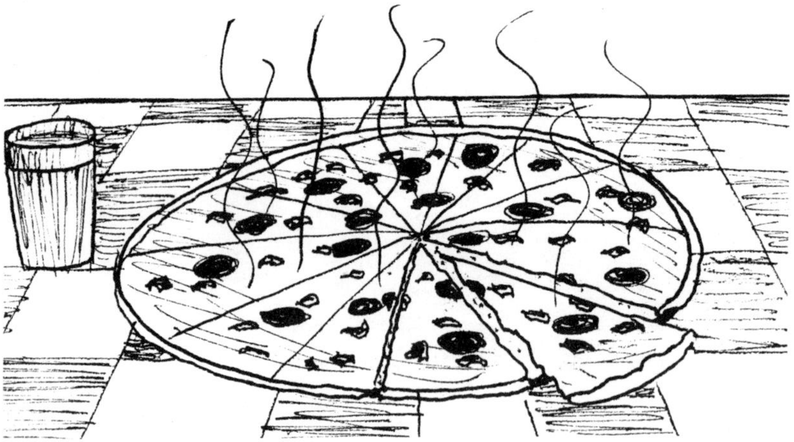

SUNDAY, AUGUST 31

After church we spend a pleasant day relaxing at home. In late afternoon we pile in the car and go for a drive, ending up at a pizza place in Hallock. After stuffing ourselves with pepperoni and sausage pizza we stop at the local park where the kids swing and play on the equipment until dark.

MONDAY, SEPTEMBER 1

The "sidewalk wetter" during the night delayed threshing. Robbie started swathing on Marion's at 10:00, while Bob and I serviced the combine.

"It's raining in Fargo," Bob announces as he pumps a grease gun. "I guess we're gonna get two inches or more."

He is prone to exaggerate but I grunt an acknowledgment of his prediction. I too have heard the weather reports and unfortunately they aren't very friendly for combining. Maybe the rain will miss us.

The sun broke through the clouds and the day took on a more promising look. At 11:00 a sample was taken which read 15.5%. At 3:00, the moisture tester read 13.5. Perfect. The 7720 paced up and down the field. The day became overcast and the humidity started to rise. By 4:00 the wheat's moisture was back to 15.5.

As the day progressed I was getting sicker with the flu. Each of the kids had been affected by it so it came as no surprise. I found myself turning up the cab's heater. At 5:00 a slight mist developed on the windshield. Robbie drove into the field with the news that he had been rained out near Kennedy. Absorbing the drizzle, the swaths ceased to yield any dust. Sending Robbie and Bob home, I intended to fill one more truck. Mud began collecting on the header's wheels. With but half a load, I was forced to quit. At home, I went directly to bed.

TUESDAY, SEPTEMBER 2

"Never pray for rain in Kittson County." The old tongue-in-cheek expression haunts me as I spend the day sitting in the house, alternately freezing and sweating. The rain continues to fall, and when it's over, 2.2 inches will be dumped at the farm and four inches on Marion's, north of Kennedy. No doubt it will be good for the beets and plowing, but combining will be miserable. Six hundred acres of unthreshed wheat remain, and it will be a struggle to finish before beet harvest. Memories of last year come to mind. Angry at the flu, weather, farming and the whole Judeo-Christian world, I bathe in self-pity while imagining the wheat swaths sprouting.

WEDNESDAY, SEPTEMBER 3

When clear skies are desired, it's unbelievable what a little sunshine can do for a man's spirits. That, accompanied by the fact that I'm feeling better, makes me feel foolish for being in such a dumpy mood yesterday. The ground was so dry and cracked, most of the water soaked in without a trace. With a little more sun we'll be combining again soon. The beets needed the rain and should yield well.

THURSDAY, SEPTEMBER 4

After three days of idleness, it felt good to be out of the house. Dislike it as I may, a trip to the local ASCS office for a commodity loan was a necessity. Past experience of standing in line for long periods has instilled a sense of dread in me whenever government business has to be conducted. Outside the courthouse a clustering of pickups gave notice of a long wait. Hoping for the best, I thought maybe people are here attending a meeting or buying license plates or paying traffic tickets or meeting with the county commissioners or . . . Come on, you idiot. Don't you suppose these farmers are doing exactly what you are? Their combines have stopped just like yours. When it's too wet for field work, what better thing is there to do but stand inside the ASCS office? Descending the wide stairs told me my second guess was correct. The large glass windows revealed a long line of farmers. My first impulse was to go home and come back another day, but I decided to gut it out. Opening the door, I was greeted by the din of typewriters. Nine or ten farmers were either standing at the counter or sitting on available chairs.

"Here comes another," one said with a smile.

"Who's on the end?" I asked.

"You're looking at him," said another. "May as well make yourself comfortable."

I found a chair and settled in for what I knew would be a long wait.

The local manager walked in and said to me, "Oh, trouble with you again?"

"Poor guy," I thought. "He can't even greet people anymore without reciting that standard salutation of his."

Keeping his eyes to the floor, he walked to his glass-enclosed office at the rear of the room. Waiting for him were a couple of high rollers whose fortunes had gone sour. Undergoing foreclosure, they seemed fidgety as the manager sat at his desk talking to them and gesturing to a stack of papers. With them was a third person I didn't recognize.

Nodding my head in the direction of the office, I asked the farmer next to me, "Who's the guy in the red tie?"

"Some darn lawyer," was his reply.

The lawyer was, characteristically, doing most of the talking. Suddenly one of the men stood up and made an animated point to the manager. As he waved his left hand his mortgaged diamond ring flashed in the light.

"Looks like things are warming up a bit," someone said under his breath, but loudly enough for all to hear.

Muffled chuckles moved throughout the waiting farmers.

Behind the counter sat a dozen desks piled with folders and papers. At each, a female clerk worked her typewriter or calculator. Taped to one desk was a beautiful picture of a jagged seashore with waves pounding on the rocks. The caption read: "Things Take Time." At another desk was a picture of four animated characters laughing hysterically. Underneath was written: "You Want It When?" Frantically dialing her telephone, a clerk momentarily listened before re-dialing. After repeating this sequence three times, she hung up. Another sat deep in thought, staring into her computer screen. As she inhaled her cigarette down to her shoes her face suddenly brightened. Bursting into action, she rapidly began poking at her keyboard. A low electronic buzz filled the room.

Getting a bit edgy, a farmer wearing a greasy John Deere cap looked at his watch and muttered, "Been here for two hours now."

The farmer sitting beside me was doing his best to be philosophical about the situation. "It's a good thing we have this place to come to for money. We sure couldn't make it with the price alone."

His understatement brought no agreement from the crowd.

"Well, I've got better things to do besides sit in this place," argued John Deere.

"They should give out numbers so we wouldn't lose our place," ventured another. Pointing in my direction, "I don't give a darn in knowing Dean Carlson's business. I could just as well come back when it's my turn."

His argument seemed sound. The conversation died and we all sat quietly with our thoughts. Attempting to explain loaning procedures to an uncooperative farmer, the clerk assigned to the counter was making little progress.

"Why in the sam hill are you doing it that way?" he challenged.

Having heard it all before, she didn't bat an eye. After a brief lecture on government regulations, he finally threw up his hands and succumbed. "Okay, anything you say."

The conversation turned to the price of commodities, in particular the poor sunflower demand.

"If I was a sunflower buyer I'd push the price down in the fall too," the philosophical farmer said. "It'll come back by spring."

Knowing he had planted a lot of sunflowers this year, I admired him for his attitude. Whether he was serious or just trying to convince himself was indecipherable.

The hands of the clock on the west wall continued to move. I killed a bit of time reading a newspaper article taped to the wall entitled "Life in the Fast Lane," which described the difficult work and high turnover in personnel at ASCS offices.

The farmer at the counter folded his papers and tucked them into the front pocket of his bib overalls. Turning to us who were left he grinned, "Next," and walked out the door.

"Okay, who's next?" demanded the clerk.

Two men stood and walked to the counter.

"What the heck are you doing?" shouted one of them. "I was here before you."

Instantly the rattle of adding machines and calculators, the beeping of computers and tapping of typewriters ceased and all eyes looked to the counter for a possible confrontation. Everything was quiet.

"Oh, sorry," the other replied, and sheepishly sat down.

Adding machines, calculators, computers and typewriters went back to work. The clock rotated. One by one the farmers moved to the counter and out the door. The man ahead of me strolled to the counter with a briefcase and opened it to expose a wealth of papers.

"This looks like a long one," I thought.

I was not to be proved wrong. Because he had many units and landlords, processing his loan took quite a while. Finally, I'm next.

"Someone should go get coffee," suggested a voice down the line. There were no volunteers. Sporadic and often separated by long periods of silence, the conversation varied from rain amounts to crop yields to how efficiency could be improved in the office. Whenever someone new would walk in, there would be the usual jokes and the same conversational topics, except with different people. After two hours and forty minutes I heard one of the clerks say those cherished words, "You're next, Dean." I sprang from my chair. Filing the previous farmer's folders, she defensively approached the counter. I tried to think of something clever to say but my mind was a blank.

"Commodity loan?" she asked.

"Yes," I replied, "wheat and barley."

"Crop share landlords?"

"Two."

"I'll TRY to explain the new procedures."

Ordinarily if someone would speak to me in that demeaning fashion I'd be insulted. But not today.

"You go right ahead," I answered. "But as soon as you see my eyes glaze over, stop."

Precisely and articulately, she explained the new regulations from Washington, D.C. Listening carefully and not daring to blink for fear that I might miss a detail, I discovered that things weren't as complicated as I had imagined. When I walked out the door ten minutes later it was my turn to smile at those still in line.

FRIDAY, SEPTEMBER 5

In September, a week's combining can be lost quickly. Lacking the summer's heat, the swaths take much longer to dry, and our unseasonably cool weather isn't helping. Highs in the 60's and lows in the 30's mean the drying process is greatly slowed.

Stepping from the pickup, I walked toward the wheat swaths. Superimposed on the wet ground, a thin veneer of slightly encrusted soil allowed my boots to remain clean. Underneath, hidden from the sun and wind, the ground remained muddy. A little sun with a stiff breeze is needed to dry these half-mile sponges. Appearing deceivingly dry on the surface, the windrowed wheat yielded wet ropes of straw upon opening.

SATURDAY, SEPTEMBER 6

Another day with no combining. Inspecting fields, I found that a couple of lighter pieces with better drainage may be fit for tillage. The barley stubble was chiseled a second time. Later in the afternoon we started to plow on Section 11. Except for one wet depression, the soil plows like warm butter. Gone are the large, hard chunks so prevalent before the rain. The heavier ground will be a different story though, and the clay soils will produce large slabs and chunks when plowed.

MONDAY, SEPTEMBER 8

Another cool sunny day. Because the standing wheat will dry quicker than the swaths, the combine was driven home and the headers switched. My hopes of threshing a few truckloads were dashed as the moisture tester read 21.6%. Using 20% as a self-imposed limit, I reluctantly parked the combine at the field's edge.

Bob continues to plow as Robbie and I dry grain and repair beet equipment. Some growers started opening beet fields today. The talk is of the nice lifting conditions. I hope it stays that way when our turn comes.

TUESDAY, SEPTEMBER 9

In spite of the clouds and cool temperature, we started combining at 9:00 north of Kennedy. With the moisture hovering at 18%, the wheat will be dried.

It was another one of those days in which the "Jobs To Do" list was longer than could be accomplished. Finally after mentally categorizing them into "Most Urgent" and "Not Quite as Urgent," I started from the top.

The most pressing job was fixing that darn dryer. Normally, when the grain temperature reaches a desired level, the spark ceases, discontinuing the flame and allowing the now-dry grain to cool. The problem was that when the spark shut off, propane would continue spraying into the combustion chamber. If a person wasn't constantly watching, much propane would be wasted while the cooling process took place. Evidently an electrical shutoff switch wasn't functioning properly. I experimented with the dryer's wiring. When bypassing switches, checking voltage, removing and replacing connectors and just plain old-fashioned, non-educated jiggling and tapping didn't work, I called Tom Gustafson, the local dealer, for help. When he came we followed the wiring diagram for discrepancies. Suddenly, there it was. The green wire from terminal A was joined to the red wire at terminal C when it should have been connected to the black wire at terminal B. No wonder. A few turns of a screwdriver solved the mystery and I was a happy man.

Two weeks to beets. Every day counts. I calculate the number of days left of combining. If the weather holds we can finish on time.

Having half the capacity of the combine, the dryer gets behind. As the afternoon progresses the 4,000 bushel "wet bin" slowly fills, and by evening it is heaped to the eaves. I drove to the field to relieve Bob. Arriving at the motionless combine, I found him working on a broken feeder chain slip clutch.

"The straw is too darn wet," he reported. "I keep slugging the feeder housing."

After a trip to the yard for parts, we used the pickup and truck headlights to provide partial illumination as Bob held a flashlight and I fumbled with a wrench. A light drizzle fell from the darkness. Dripping off my cap, it seemed to intensify.

"Looks like it's getting harder," Bob observed.

Grunting an agreement, I struggled with the stubborn slip clutch. The cold night air was penetrating my wet jacket. As we worked directly beneath the idling engine, directions and communication had to be shouted. Finally, with all the parts intact and tightened, the slip clutch appeared operable. Gathering tools by flashlight, we stood in the glare of the truck's headlights. Lighting a cigarette, Bob gazed into the cloudy darkness.

"I can keep on," he said.

"No," I replied. "You go home and get some sleep. I'll fill trucks."

"Might be too wet."

"Maybe."

"How 'bout tomorrow?"

"I'll call if it rains too much. Otherwise plan on coming at 8:00."

"Yeah, I don't know what it's supposed to do." We both leaned against the truck while he finished his smoke. Tired from my busy day, I contemplated quitting. The rain continued to fall in a light drizzle.

"Well, I'll give it a whirl and see what happens," I said, and walked to the combine.

Gathering his lunch kit, Bob drove off in the pickup. Climbing into the cab, I removed my wet jacket and turned up the heat. The rain had lightened somewhat. Swinging the combine into position, I opened the throttle and engaged the separator. Moving eastward, the disappearing taillights of Bob's pickup were visible. Three miles east, four blond youngsters were going to bed.

"Dad, do you have a copy?" the oldest calls over the CB.

"Yes, I do, Melissa."

" I just wanted to say good night."

"Good night, Melissa."

"Good night, Dad."

"Good night, Alan."

"Good night, Dad."

"Good night, Todd."

"If Andrew could talk, he'd like to say good night too."

"Tell him good night for me."

"10-4."

Traveling slower than usual, the combine groans as it eats the tough, wet straw. Chunks are thrown through the chopper. Cars passing on highway 75 reveal a wet pavement. Eventually the pile of wheat in the truck box appears, and with one more hopper it is filled. Debating on whether or not to fill another, I check the moisture. 20%. My watch shows midnight. Seventeen hours is enough for one day.

WEDNESDAY, SEPTEMBER 10

It rained just enough to halt combining, so Robbie and I spend the day preparing the equipment for beet harvest and drying grain. The cool, humid day slowed the drying process.

At least Bob is able to continue plowing. Having just refueled the Versatile, I watched as it drove away. Drawing eight furrows at a time, the plow cut long, black slabs of the old lake bed.

A familiar pickup stopped; its driver leaned out the window. "Seems like it's either too wet or too dry," he said, referring to the plowing.

Laughing, I replied, "I don't know when this land ever plows decent."

"When do you open beets?"

"24th."

"Two days before us. I hate these late starts."

In the rotational process, a beet grower's turn to open fields prior to the main campaign is varied from year to year.

"Chances are we won't get all our quota in," he added.

"Think you'll be done combining by then?" I asked.

"I sure hope so. We've combined and dug beets at the same time other years and it's no fun."

THURSDAY, SEPTEMBER 11

A bright sunny day. Having straight combined all that's possible, Robbie began swathing. While Bob plowed, I combined on Marion's. Since the moisture content has fallen to 15%, the wheat can be binned, provided it has proper aeration. By 4:00 my first truck load of the day was filled. Robbie arrived from swathing and hauled grain the rest of the day.

Combining along a major highway like U.S. 75 provides company. Traveling from somewhere to someplace, the assortment of trucks, pickups and cars supplies a constantly changing scene. I recognize some vehicles, but the majority are driven by strangers. A long Burlington Northern freight train passes slowly. From an open locomotive window the engineer waves. I count the boxcars. A friend of mine who sells insurance parks his Ford Bronco at the crossing and squeezes into my cab with me. We ride a couple of rounds together while chitchatting about farming, football and golf.

Under the swaths, the ground is still wet. Occasionally mud must be cleaned from the feeder housing's throat. The wet straw next to the ground dictates a slower ground speed than desired, but if we continue to plod along, we'll get done.

FRIDAY, SEPTEMBER 12

"If this keeps up we just might get done before beets," exclaims Robbie as he gets out of his car.

"Keep your fingers crossed," I advised.

The mercury dipped to 27 degrees this morning, the first major killing frost. After giving directions to Robbie and Bob, I drove to the combine. The morning air was cool as I struggled into my coveralls.

Overhead, all by himself, a goose was flying south.

"Can't say as I blame him," I said to myself.

Realizing that the best part of the year is over, everyone engages in the same conversation. Where has summer gone? Those who wish they had enjoyed more relaxation or activities in its brief span are complaining. Fueling the combine, I savor the good times and memories of hot sun. A pickup turns off the highway and into the field. I recognize the driver as a custom combiner.

"Morning," he smiles. "Heard you might be needing some help."

"If the grain was dry, I probably could," I reply. "What are you charging?"

"$14.50 and you furnish the fuel."

"My problem is that the dryer could never keep up. We'd have the farm full of wet wheat, then have to move everything out to dry."

"Yeah, I know what you mean."

We talk for awhile and he leaves his card. As he drives away I second-guess myself, thinking that I should hire a custom combiner. However, after today only the poor crop will be left, and to give seven bushels away on a crop that's yielding fifteen to twenty doesn't make sense.

Robbie continues to swath while Bob and I combine on Marion's south quarter. We switch jobs for a couple of hours in the afternoon, and while I'm dumping a truck at the bin site adjacent to the highway, a small, generic, rental-type car stops at the crossing. A lady armed with a camera and a big lens steps from the passenger side. Standing beside the uplifted box and grain auger, I realize my picture is about to be taken. Making sure my shirt is tucked in and straightening my posture, I give my best "farmer pose." Aiming in my direction, she snaps the shutter. After we exchange waves the unknown photographer and her driver disappear down the highway. Wonder creeps over me. Who could that have been? Why did she want a picture? She must have been a professional photographer. Of course! It's a *National Geographic* team! They flew to Grand Forks, rented a car and are driving the countryside looking for a feature story. Imagine that! A full-page color picture of me in *National Geographic*. No doubt the caption will read "Hard Working Farmer Racing Against Nature to Harvest His Wheat Crop."

By late afternoon we finish by Kennedy and move to Hallock. Contrasting with the forty- to forty-five bushel crop just harvested, the crop on Dad's and Clara's is downright lousy. At midnight, I climb off the silent combine. A seventeen hour day.

SATURDAY, SEPTEMBER 13

The days spent combining become blurred. My mind is preoccupied with the tasks at hand. I rush from job to job. If repairs are needed they are hurriedly done. Obsessed with combining as many acres as possible, I fail to notice ordinary things about me. Receiving so many bits of information, our minds automatically push those seemingly unimportant details into our subconscious without proper digestion. Each night while driving home, I reflect upon the day and discover how many things I've seen and recall what I've said. I regret not taking a minute or two here and there to explore some thought. If I weren't keeping a journal of the day's activities, I would forget entirely what I did. Days are categorized into "good" or "poor" depending on how many acres are covered. I get up every morning, give directions to the men, combine all day until midnight and get to bed in the early morning hours. I look forward to Sundays for a little rest and time with my family.

Today was another blur, a repetition of the previous day. One of the highlights is when Carol and the four kids bring afternoon lunch to the field. I let Bob take the combine while I get a chance to see my family. The kids act insane. They're in the truck box, rolling in the grain. Throwing dirt chunks at one another, they race up and down the field. Everyone is talking at once telling me of their day's activities. Finally, I have to make only one speak at a time. The homemade rolls and freshly-baked cake are delicious. When lunch break is over, I catch the combine as it drives past, climb its steps, open the cab's door and slide into the seat as Bob steps out. Without slowing down, the switch is made. Bob takes the fully loaded truck home.

As night approaches while I'm combining on Clara's 60, a flock of ducks circles Dad's pond and glides to the water. Yard lights come on and house lights extinguish. Lights in the east denote either Harold Haugstad or John Anderson trying to finish combining too. At midnight the twin screws are filled. Indecision plagues me as I contemplate continuing another hour to fill the tag axle. A small mental voice tells me that I'm too tired, while another reminds me of the long winter in which I can catch up on sleep. Listening to the former, I pull the fuel shutoff. In the distance, the lights of the neighbors' combines give notice of their physical stamina. Bed will feel good tonight.

MONDAY, SEPTEMBER 15

Yesterday's break and the good night's sleep put both the hired men and me in good spirits as we discussed the day's activities.

"I want you to keep on that plow," I told Bob.

"Sounds good to me," he replied. "We've got to be cracking to finish before beets."

He crisply walked to the service pickup and sped out of the yard in the direction of the waiting 850 Versatile. Robbie had gathered his necessary tools for the day's swathing.

"Should be done by noon," he said.

"Fine," I replied. "I'll have the trucks filled for you by that time."

Robbie and Bob have the same sense of urgency about getting the work done as I do, and I feel fortunate in having two excellent hired men.

The day was everything I hoped it would be. We unloaded on the go without losing a minute of precious time, and the acres disappeared quickly. With the south wind drying the swaths, we were able to bin the wheat directly from the field. Calculating the number of days required for combining, plowing, discing and anhydrous ammonia application, I realize we won't be able to have everything completed before beet harvest. I hope we can secure enough good help to do beets and tillage at the same time.

The day was going a little too smoothly. At 7:00 the feeder chain slip clutch mechanism broke again. After running home for the proper tools and acetylene torch, Robbie and I fixed it by flashlight.

"This is getting a little old," he complained.

The acetylene torch illuminated the whole front of the combine as it cleanly cut the worn slip clutch free. Completing the removal at 9:30, we realized we were done for the day. A new slip clutch would have to be purchased in the morning.

"Looks like we lost a good five hours," I said.

"Could've gone all night with this wind," Robbie answered. Stepping forward, he spat a wad of chewing tobacco. "How's the forecast?"

"Not good." I walked away from the combine to check the wind direction. The air was cold. I zipped my jacket to the neck.

"East wind," I announced. "We'll just have to give her heck tomorrow and get what we can before it rains."

TUESDAY, SEPTEMBER 16

"Seventy percent chance of rain," exclaimed Bob, walking into the Quonset. Considering the red morning sky and the east wind blowing all night, I had to agree.

Parts were secured from John Deere and the slip clutch was fixed in record time. A quick dusting of the windshield substituted for the usually meticulous servicing of the combine. Expressing both our sentiments, Robbie patted the combine's tire saying, "Sorry old girl, but that's all we've got time for today."

The skies became overcast, and at 2:00 the rain began. We halted combining and Bob continued plowing.

WEDNESDAY, SEPTEMBER 17 TO SATURDAY, SEPTEMBER 20

Carbon copies of one another, these four days are classified as problem solvers. For instance: Bob drives into the yard with the news that the front wheel of the plow has broken off. Loading a jack and tools into the pickup, we take a slow, jarring trip to the middle of a plowed field, remove the framework of the wheel and bring it to the shop. A half-hour later after some heating, welding and a little John Deere green paint, it looks as good as new. The rough journey is repeated and the plow is reassembled.

A hydraulic cylinder on the beet harvester is leaking. It's disassembled and the O rings are replaced. The lights on one of the trucks won't work and a break in the wiring is found. Working on job after job, Robbie and I repair beet equipment and trucks in anticipation of next week's beet harvest.

Possessing a certain mystique, the demon of mechanical breakdown lurks everywhere during beet lifting. Aggravated by muddy working conditions, a bearing's life is slashed to a fraction of its normal span. Trucks that would never see the inside of a repair shop find themselves with twisted drive shafts, blown engines or ruined differentials. Practicing preventive maintenance means bearings, chain and flails are indiscriminately replaced. Going through this annual ritual, the equipment arrives at the point where the known weaknesses have been fixed. Whatever else breaks will have to be fixed during the campaign.

At the end of each day I'm exhausted. A soft chair in the evening feels good.

The weather has also been a carbon copy each day with cloudiness and a light drizzle falling periodically.

MONDAY, SEPTEMBER 22

"This has got to be one of the cloudiest Septembers I can remember," comments a farmer while eating his fried egg.

"It's supposed to break this afternoon," replies another.

"We need some sun for those beets. I doubt if they've put on much tonnage this month."

"I don't know where summer has gone."

The breakfast crowd's conversation reflected a melancholy mood. Summer has gone, and there is a change in the air. Having shed its fuzzy seeds, the Canada thistle is brown with dried pods where its purple blossoms once bloomed. The goldenrod has turned from yellow to brown. Congregating in anticipation of the southward journey, the meadowlarks will soon be abandoning us for warmer temperatures.

After a seemingly endless string of cloudy days, the sun broke through this afternoon. Last night's half-inch of rain relegated us to yard work. Looking up from a truck engine, I saw a brightly-colored pickup drive into the yard. Recognizing the driver as a familiar custom combiner, I approached his vehicle.

"How ya doin'?" he drawled in his Oklahoma accent.

"Fine. How's the run been?"

"Oh, pretty good," he replied, putting a pinch of tobacco in his cheek. Offering the box to me, he asked, "Chew?"

"No thanks, I'm trying to quit." He laughed loudly at my canned response.

"Guess I'm about the only custom cutter in these parts. Everyone else has split."

"Can't say as I blame them."

"I called home this morning. It was 80 degrees there yesterday."

"I could handle that," I laughed.

After talking a few minutes I explained I didn't need any help but to look me up next year. Promising he would, he climbed into his pickup and drove off with a friendly wave.

The temperature rose to a beautiful 70 degrees. At 4:00 I drove the rotobeeter to Freda's quarter and started to top beets. After so many days of fixing, it felt good to sit in a tractor again. The 4020 John Deere groaned as the rotobeeter's rotating flails chopped and tore the tall green foliage. After the leaves are removed, the trailing hydraulic scalpers neatly trim the beet's crown. Denuded of their leaves and petioles, the newly-topped sugar beets protrude above the freshly-created wet silage. Four months ago today these beets were planted.

TUESDAY, SEPTEMBER 23

The westerly breeze and 70-degree temperature gave us hopes of combining. However, after threshing a sample at 4:00 and finding the wheat at 21.6% moisture, I parked the combine. With sixty acres of wheat remaining in the swath on Clara's east quarter and twenty acres on Dad's quarter, we'll either have to combine after beets, or, if enough help can be found, during beet harvest. Rotobeeting all day, by late afternoon Robbie had completed the beet field's headlands and had split the quarter into half a dozen smaller fields. Tomorrow we'll start lifting.

WEDNESDAY, SEPTEMBER 24

Beet lifting has finally begun. Pulling the beet cart with the 4020, Robbie drives alongside the lifter. Beets are planted in rows spaced at twenty-two inches and require that the tractor's and cart's wheels are spaced at eighty-eight inches, thereby straddling four rows. Utilizing a beet cart instead of loading directly into a truck when opening fields does far less damage to the beets. A cart will straddle the yet undug rows but a truck will squash them into the ground. Holding seven tons of beets when it's full, the cart's contents are elevated into a truck. After a couple of rounds, enough beets are dug to enable the truck to drive on ground already worked and to be loaded directly from the harvester.

Some people have described the beet lifter as resembling a long-necked dinosaur. Harvesting four rows at a time, rotating lifter wheels pinch and pull the beets from the ground. I'm amazed at how well it lifts. Paddles knock the newly dug beets, rolling and jumping, backward onto rapidly spinning grab rollers. It is here that most of the dirt is removed from the beet. Equipped with raised edges, the grab rollers tear at the beets, dislodging mud and dirt as they are moved toward the unloading conveyor and up into the waiting truck.

The dirt is falling off the lifter wheels without a bit of mud. From the 4020 pulling the cart, Robbie smiles and gives a thumbs-up gesture. I know what he means. After last year's wet harvest, we're appreciative of dry digging conditions. The twin screws are loaded and disappear down the road, their boxes heaped with the white beets.

One-and-a-half hours later, a distant truck appears in the south. Recognizing the white twin screw, I warm the tractor. Arriving at the field, Bob dumps his dirt and swings under the waiting boom. Robbie arrives with the red twin as Bob and I are loading. When we've driven a quarter-mile down the field, a rain shower hits from the southwest. Continuing as long as possible, the truck finally loses its traction on the slippery surface. Watching from a distance, Robbie realizes our plight and races the Versatile to our rescue. He backs it to the truck's front. A tow rope is attached to the bumper and the truck gets loaded while being pulled alongside the lifter. When it's full, I wave them to the crossing.

As the truck pulls up the crossing, I realize they are taking the corner too sharply. The truck's left rear wheels slip off the crossing's edge. From the truck Bob waves for Robbie to stop. Halting, the truck heaped with beets sits precariously at such an angle that I wonder how it keeps from falling on its side. Standing in the rain and analyzing our potentially dangerous situation, we decide to ease the truck back down the now-slippery crossing for another attempt.

"Lock the door and close the window," I advise Bob. "In case she tips, don't worry. My workman's compensation is paid." He didn't appreciate my humor.

Motioning while the rain penetrated my jacket, I directed traffic. The tractor and truck backed up in unison. The truck righted itself and slid down the crossing. A wider angle was plotted on the second attempt, which brought the truck safely to the road. Our first day of beet lifting is short.

THURSDAY, SEPTEMBER 25

Too wet to plow. Insulating the ground, the wet mat of straw hinders drying. Attempting to aerate and enhance drying, Bob drags a chisel plow across the remaining eighty on Clara's east quarter.

At least the sun is shining. I hope we don't get a big soaker. Water sits on the beet field's headlands. After dinner, Robbie and I manage to load two trucks. Using the cart, we dig until the water holes are reached before turning back. Perhaps lifting in this fashion doesn't make much sense. But in a game where every load counts, two loads are better than nothing. We arrive at the Drayton beet plant and the sign on the scale house reads: "Scales Close at 6:00." That's okay, it's too wet to lift anyway. With so few trucks coming, only one piler is operating. Seven others wait silently for the main harvest. Filling the air, the aroma of burnt sugar emits from the huge mill.

Driving onto the piler's platform, I raise the truck's hoist to dump its load. As the box rises, the endgate pops open. The truck shakes as seventeen tons of sugar beets rumble into the piler's hopper and are transported up the long conveyor into the growing mountain of brown beets. Cleaning the beets further, the piler collects dirt which is elevated into my now-empty truck. This dirt will be dumped back onto the field. As I weigh my empty truck at the scale house, I feel that lifting should go better tomorrow.

FRIDAY, SEPTEMBER 26

Having engaged themselves to a potato farmer for the duration of his harvest, two of our beet drivers can't come until Monday. Calling the Grafton unemployment office a few days ago, I left my name, address and type of employee needed. In the heart of a busy harvest season, I should have realized that all the good men have already found jobs.

When the gold Monte Carlo drove into the yard I met Bret for the first time. Grinning, a short, young man with wire-rimmed glasses stepped from the car.

"Grafton Job Service said you were looking for a beet truck driver," he said politely.

"Yes, I am. Who have you been working for?" I asked.

"Altendorfs, from Grafton."

"Did you drive truck for them?"

"Yeah, I drove lots of truck for them."

"Why'd you quit?"

"I couldn't take any more of the verbal abuse. Man, they would just yell at a guy. Finally a buddy of mine and I just quit."

His last statement should have been a clue as to his driving ability. During the next few days I would deeply regret not calling Altendorfs for a reference. Bret immediately became the harvest crew's weak link. If there was a hole in which to get stuck, Bret would find it. Sitting with his truck mired in the mud, he would senselessly roar the engine and spin the tires until another driver would walk over and tell him to stop. I thought that eventually he would learn that he can't drive through the same holes time and again. But not Bret.

Trying desperately to be Mister Nice Guy, I wouldn't yell at him whenever he'd run his truck into the beet cart's unloading conveyor. Politely reprimanding him, I thought he would catch on. But not Bret. Even a totally inexperienced driver would, after a load or two, comprehend the basic routine of beet driving. But not Bret. Waving frantically from the tractor cab, I felt my blood pressure rise as he would constantly rub his truck box against the lifter's unloading boom. Finally looking in my direction, he would always overreact by driving so far away the beets would completely miss the truck. Looking at me with his cheesy smile, he was fully unaware of the trail of beets falling from the side of the truck. Continually stopping the tractor to give him instructions, I found my sentences were becoming more terse.

In my yard, developing from the recent rains, was a wet hole from which the drivers were instructed to stay away. Upon completion of each shift, the boys would park the loaded trucks on the firm, high ground of the yard. Bret, however, would manage to bury his in the hole. When the Altendorfs reach the Pearly Gates I hope Saint Peter gives them stars in their crowns for limiting their abuse of Bret to verbal. I was tempted to do much more.

Lifting wasn't going too badly. The two twin screws were driving out of the field on their own power, but the tag axle required pulling. A small rock caught between two grab rollers on the lifter, snapping three V-belts. Taking five pairs of hands two hours, the repair work was completed just as another shower struck. It didn't last very long and we were able to continue lifting.

The white twin screw has been plagued by an overheating problem. It first appeared during grain harvest, and the local mechanics have been employing the trial-and-error-and-you-pay-the-bill method in solving the dilemma. The radiator has been boiled out, thermostats replaced, timing adjusted and water pump exchanged. Still the engine boils over. Finally by late afternoon, after loading three tons of beets into its box, the 427 Chevy could no longer contain itself. Stopping the lifter, we watched helplessly as it emptied its green coolant onto the muddy ground. I explained the situation to the local Chevrolet garage's shop foreman and he said, "Okay, bring it in. I'll have two guys tear it apart right away and we'll put heads on it."

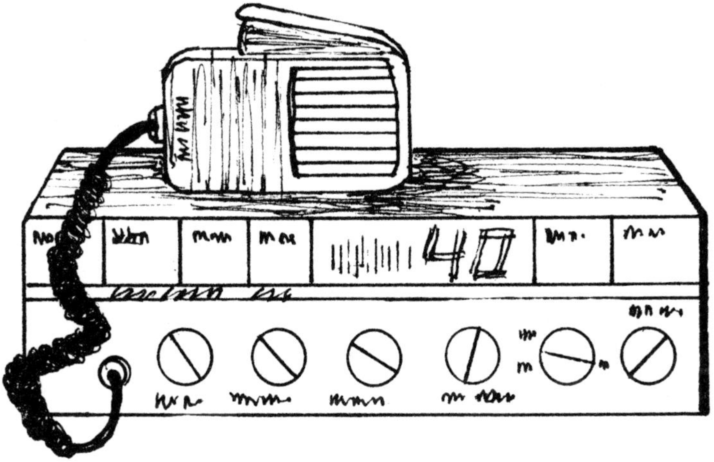

SATURDAY, SEPTEMBER 27

While the white twin was in the shop, we loaded beets with only two trucks. At 2:00 the garage called, informing us the heads were replaced. Bob drove it to the field from Kennedy. Without looking in the box, he raised the hoist to dump what he thought was dirt. Standing at the lifter a quarter of a mile away, I saw what was happening. Scrambling to the 4430 and grabbing the CB, I shouted, "Bob, don't raise the hoist!"

Too late. The endgate sprang open and yesterday's freshly dug beets poured from the truck. I thought of the work I had invested planting, cultivating, spraying, thinning, rotobeeting and finally safely lifting them to a truck, only to have them dumped on the ground. I was disgusted. Why didn't he check the box before dumping? Bob got out of the truck and stared for a few seconds at the pile. I knew what was going through his mind. Driving his now-empty truck to the lifter, he looked pretty sheepish. Getting out, he walked to the tractor and said, "Do you want to kick me where the sun don't shine?"

Trying to keep from laughing, I was doing my best to act really mad.

"If I'd had a shotgun, I'd have blown your rear end away."

Seeing through my charade, he laughed, "That's probably only the second mistake I've made."

I didn't ask what the first was. We proceeded to dig the load. After loading the trucks, I was cleaning mud from the lifter when I heard my old friends. High overhead in a southward facing V-formation, the geese were migrating. As the sight of them in the spring brings happiness, watching them wing their way to their winter homes puts me in a somewhat dismal mood.

"Have a good winter," I said to them as they continued out of sight over the southern horizon.

A letter from Crystal Sugar informs me we are to start lifting full time on Monday instead of Wednesday as originally planned. In order to be ready it looks as if we'll have to work tomorrow.

SUNDAY, SEPTEMBER 28

For a mixture of religious and personal reasons I've always regarded Sundays as a special day of attending church, relaxing and playing with the kids. After six days of physical and mental exertion, a day of idleness is perfect therapy. Today was not to be such a day. If beets don't get topped, we won't be able to start lifting tomorrow. This morning I put on my same old dirty coveralls, donned the same dirty cap with the grease-stained brim, laced the same work boots and slipped into the same old mud-caked gloves before walking out the door. Usually a special day, Sunday was relegated to just another ordinary work day. My glum mood wasn't helped when my eight-year-old daughter, Melissa, informed me that the Bible says you shouldn't work on Sunday. Be that as it may, enough beets were rotobeeted for a full day's work tomorrow.

MONDAY, SEPTEMBER 29

The big day finally came. What has been called the most dramatic movement of a crop to market anywhere in the world has begun. Up and down the Red River Valley, 1,700 sugar beet growers will deliver up to 35,000 truckloads of beets per day to the receiving stations. Trying to deliver as many loads per day as possible, the growers fill their trucks to the legal limit, race to the pilers, dump their beets and hurry back to the field. Safety is continually stressed but it's an unusual year that doesn't have some truck tipped on its side, its cargo of beets spilled into the ditch. Whether one raises beets or not, it's the topic of conversation everywhere. Older ladies whisper in church of the harvest's progress.

At the Hartz store and coffee shop, men complain, "I was driving sixty miles per hour and Peterson's* truck passed me like I was standing still."

"It's crazy. I stay off the road this time of year."

To our full crew I briefly described the trucks they were to drive. "The object," I facetiously explained, "is to get to Drayton and back as quickly as possible. If your mother gets in the way, run her over, too." Experienced drivers all, they know what to do. Bret, of course, is the exception.

American Crystal receives beets twenty-four hours a day and assigns shifts to each grower during which he can deliver his crop.

Since this year our shift is 11:00 p.m. to 11:00 a.m., sleep will be at a premium. Today will be a short day: after the scales open at 8:00 a.m., we'll only be able to haul until 11:00. Arriving here at 7:00, the drivers had the previously-loaded trucks at Drayton by 8:00.

Hauling twenty-six miles from the fields to the mill and back takes time. Under ideal conditions the round trip takes one-and-a-half hours.

Back at the field, the trucks were loaded again and sent on their way. Another trip was not possible before the shift ended. The first two trucks were loaded and sent to the yard awaiting tomorrow's (tonight's) shift. While we were loading the third, rain began to fall. Before it stopped, a half-inch would accumulate. Now we really are wet. Too bad. Lifting conditions were ideal. From now on we can expect to pull trucks and clean lots of mud.

At 6:00 p.m. I called the drivers, telling them not to come to work tonight. The pilers have shut down, awaiting dry conditions.

TUESDAY, SEPTEMBER 30

Going off and on shift is a tiring experience. A person's body could possibly get accustomed to working all night, and a sleep routine could be developed if it weren't for the weather interruptions.

Bret drove into the yard at 1:00 this afternoon and told me he was quitting.

"I hope you're not too upset," he said.

"Well, of course I'm upset!" I shouted. "When anyone quits on me in the middle of beet harvest it makes me upset!"

I grabbed my checkbook and scribbled out enough money to pay him in full. All the while I was inwardly jumping for joy at his resignation. My performance was worthy of an Academy Award. In my mind's eye I envisioned myself humbly accepting the Oscar for best actor. To the standing ovation of thousands, I would lovingly hold the coveted trophy and tearfully say a few words of thanks to all the people who made this possible.

"Here , take it and go," I snarled.

Sneaking to his car, he apologized, "I'm really sorry."

Watching him drive from the yard, I momentarily regretted being so cruel. The feeling soon passed, and I spent the rest of the day savoring my theatrics.

The day is bright and sunny. I've driven to the beet field three times, checking for dryness. Some growers on sandier land started to dig by late afternoon. The pilers are "off shift," meaning that anyone who is able to dig can haul, regardless of his shift. Tomorrow we will try.

WEDNESDAY, OCTOBER 1

Arriving at 6:00 this morning, the drivers and I started loading beets in the dark. Dawn revealed a cold, cloudy day. Having anticipated pulling the trucks right from the start, I was surprised at how well lifting was going. Driving under their own power, the trucks could obtain half a load before bogging down in the mud. Watching from the Versatile and anticipating where the truck would stall, Robbie would be only a few moments away. Priding himself in his efficiency, each driver would step from the truck cab, grab the rope dragging from the tractor and quickly attach it to his truck's front bumper.

Communicating through a series of hand signals, Robbie would pull the partially loaded truck alongside the lifter. Having repeated this procedure a hundred times in previous years, he would set his tractor rpm's to correspond with the lifter tractor's ground speed. If the truck needed to be pulled onto the road, he did that also. A loaded truck on the road driving toward the distant mill is a welcome sight.

"That's one less," I say to myself.

The beet plant is still off shift. We dig all day and into the night. About 9:30 another rain came, forcing us to quit.

THURSDAY, OCTOBER 2

Last night's rain has made lifting impossible. Today's weather is a repeat of yesterday. Walking into the ASCS office in Hallock, I was told that loans aren't made until 1:00. Returning after dinner and finding seventeen farmers ahead of me, I went home. The rest of the day was spent fixing a bent power takeoff shaft on one of the trucks.

FRIDAY, OCTOBER 3

Still too wet to lift. I hope it doesn't turn out like last fall. The weather is cool with an occasional shower. We'll just have to wait until it's fit. Determined not to stand in line, I arrived at the ASCS office at 8:00 and shortly after, the place was full of farmers wanting to take commodity loans.

SATURDAY, OCTOBER 4

Still no lifting. With beet harvest, combining, tillage and anhydrous work left to be done, we need some Indian summer weather. I caught myself wishing that October was over and all the farm work completed. Keeping warm in the house, I could relax and watch the first snowfall gently cover the ground. It seems as though this cycle of rainy weather will have to run its course, as it always does.

There'll be plenty of sunny days yet this month in which to finish beets and combining. Sitting around and waiting for them is the tough part. It's a dirty-looking time of year. The birds are gone. Outdoor recreational activities are limited. A round of golf would be great, but it's too cold and miserable for that, too.

Parking a hundred yards from the road on Freda's quarter, I slowly walk to the stationary lifter. Underfoot, the dark slippery ground is patched with slowly-growing areas of gray. If the sun comes out we might try tomorrow. I walk around the lifter, kick the tires and clean some dried dirt off the grab rollers. The sharp northwest wind is a good sign.

SUNDAY, OCTOBER 5

The day dawned sunny. I called the drivers to try lifting at noon. Church services consisted of the television ministry. Arriving at 11:30, the men and I were loading beets by noon. The two twin screws can run under their own power. The tag axle is another story. Whenever I see it return from Drayton, I look for the Versatile, knowing it will have to be pulled.

Unlike any other type of field work, days are measured in loads, not acres. Every load driving down the dustless township road is another fruit of the day's labor.

Tire cleaners installed on two of the trucks cause most mud to be left in the field. One truck, however, upon reaching the road with its tires filled with mud, leaves a trail of dirt from the crossing.

Getting out of the tractor three or four times a load to clean mud from the harvester makes lifting slow. A little piece of protruding metal or obstruction on the lifter is a great place for mud to accumulate. Slowing or even halting the movement of beets, the mud must be removed the old fashioned way: with screwdriver, crowbar and scraper. Pulling the beets from the ground, the lifter wheels will occasionally plug, requiring cleaning.

As the afternoon progresses, the warm sunshine gradually improves conditions. Anticipating that all growers will begin digging on shift, I listen for the broadcast beet harvest report. Hearing the usual preliminary Cenex commercial, I turn up the volume on the tractor's radio.

"This is the Drayton factory beet harvest report for Sunday, October 5th. We are still off shift. Repeat. We are still off shift. Those growers who are able to dig may do so." After giving figures as to the tons sliced yesterday and percent sugar, the announcer continues, "Stay tuned for further developments."

With a truckload per round, an eighteen-ton yield is feasible. Parked on the headland waiting for the trucks to return and scraping mud from the lifter, I looked up to see Glen Gunnarson's pickup drive down the crossing. Casting a knowing eye over the lifter he said, "It's a little muddy, isn't it?"

"Yes," I nodded. "We're taking it a load at a time. Are you lifting?"

"Not today. We'll go tomorrow."

Sitting in his warm pickup, he pointed a pack of cigarettes in my direction.

"Smoke?"

Having quit smoking ten years ago, I was a bit apprehensive. Momentarily hesitating, I capitulated to his offer.

"I hate beet harvest," he said, lighting one for himself. "If it's dry, it's kinda fun. But when it's wet . . ." His voice trailed off as if he were afraid to mention the extra work and expense.

Two miles away, the red twin screw raced in our direction. Seeing it approach, he said, "Here's your truck. Guess I'll be going. Good luck."

Loading each truck, the usual dialogue takes place.

"How are things in the yard?" I ask over the CB.

"Good, only two trucks ahead of me."

"What piler did you go to?"

"Number three."

"Anybody ever say if they were going back on shift?"

"Not that I heard."

We managed eleven loads in ten hours before one of the drivers came from Drayton with the expected news.

"They're going on shift."

Baring my watch to a truck's headlights, I saw it was 10:00. Since the beet plant was going on shift our normal 11:00 p.m. shift would just be starting. Standing in the glaring light of the lifter tractor, the drivers and I discussed our situation.

"I hate to ask this of you," I said, "but could you go home and get a little sleep and be back at 2:00?"

Looking at his watch, Bob did some mental calculating. "We should be able to get a couple hours' sleep. Sure, I can do that."

I apologetically looked at the other two.

"No problem," replied one.

"I can be here," said the other.

They're a gutsy bunch. I guess they realize that in beet harvest, sometimes sleep is hard to come by. With one truck already full, we loaded the remaining two. When the last one wasn't filled until 11:30, I told its driver not to come back until 3:00.

MONDAY, OCTOBER 6

The one-and-a-half hours' sleep wasn't enough, but my high-strung emotions pop me out of bed. Two of the drivers arrive at 2:00 a.m. and within a few minutes, their red clearance lights are disappearing into the moonless night. I drive to the beet field and shine the pickup's lights on the lifter. After sipping a half-cup of hot coffee and listening to an all-night talk show host solve all his callers' problems, I step into the cool night air. Mud is everywhere. It sticks to your boots and coveralls as you lie in it, cleaning lifter wheels and grab rollers. It permeates through three layers of clothing, and your knees feel cold and wet. Its sticky mass renders your gloves totally unworkable. You adjust your cap, it gets on the brim. It hides in every nook and corner of the beet lifter, collects on the lifter tractor's floor and seat, fills the tractor's duals, coats roller chains and bearings and camouflages the trucks. At coffee time, the lunch kit and thermos bottle get smeared. The curse of beet harvest, it's the source of much physical fatigue and mechanical breakdown.

While cleaning the lifter, I occasionally look toward the west in anticipation of headlights. Seeing a set in the distance, I warm up the tractor. By 3:30 we are digging. In spite of the fact that many stops must be made to clean, lifting is going pretty well. We dig into the dawn. Finally by 1:00 p.m. the last truckload is lifted and the truck parked in the yard, awaiting the coming evening's shift. After fueling the tractor and greasing the lifter, I drive home.

I unwind with the daily newspaper, eat, (breakfast, dinner or supper, I don't know which,) and try to get some sleep. Hiding the midafternoon sun, the black garbage bags draped over windows darken the bedroom, but sleep does not come easily. My body wants to rest but my mind is still digging beets. I manage to get three hours.

At 10:15 the drivers arrive. The pre-loaded trucks are serviced and warmed before making their journey to Drayton. Crossing the scales at 11:00, the first one gets back to the field at midnight.

TUESDAY, OCTOBER 7

Night and day become irrelevant. We bathe the lifter in artificial light as the digging continues. We average three stops per round to clean mud, and the trucks are slowly loaded. My arms ache from scraping. Before long, a routine is established. Stopping, I grab a crowbar, crawl on top of the grab rollers and commence cleaning while the truck driver uses another on the wheels. When that is accomplished, a long-handled scraper is used for flatter surfaces.

The air is cool and the conversation sparse. Occasionally the mud gets cursed. Relieved when their trucks are full, the drivers can remain in their warm cabs for the next one-and-a-half to two hours.

A gray eastern sky was developing when another rain shower forced us to quit. After going home and catching some sleep, I drove to the field and cleaned the tractor's duals. Now the sun is shining, making lifting tonight possible. I call the drivers, telling them to come at the usual time.

WEDNESDAY, OCTOBER 8

Sleep is a commodity to be cherished. All night we dug in the mud. At times it seemed as though we would have to quit, but the trucks kept getting loaded. My legs and arms grow weary from climbing and scraping. Toward the end of each load, the trucks must be pulled.

Preceding a cold front, an unseasonable thundershower came crashing from the northwest at 4:00 a.m., its lightning bolts flashing in the sky around us. Dropping only a small amount of rain, it nevertheless soaked our jackets while we chipped away at the mud with crowbars and scrapers. Somehow, cleaning mud in the middle of the night in a rain shower tarnished the glamour of raising beets. After the shower, the wind went to the northwest and the air cooled dramatically. Toward morning, one of the drivers wanted to sit in the truck while I cleaned the lifter. I politely asked him to give me a hand.

"You're going to make me work for my six dollars," he smiled.

"That's right," I answered, not looking up from my job.

By 2:00 this afternoon we had managed to dig seventeen loads. Completely exhausted, I went home and got a luxurious three-and-half-hours' sleep.

"Dad, could you read us a story?" Alan asks.

It is 7:00 p.m. and I'm sitting in the living room, trying to wake up. Since the kids have an 8:00 bedtime, I'll see them exactly one hour today.

"Sure," I answer groggily, "you find a book."

Melissa, Alan, Todd and I sit on the couch and I read aloud "The Fox Went Out on a Chilly Night," "A Thousand Pails of Water," "Harry the Dirty Dog," and other short stories from the *World Book Encyclopedia's Childcraft*. I'm glad that I'm able to spend at least a little time with the kids, reading them stories between beet harvest shifts.

It's 11:15 p.m. The first truck should be back from Drayton in half an hour. Carefully avoiding the deep ruts, I drive down the field, searching for the lifter. Sitting where it was left ten hours ago, it appears in the pickup's headlights. I park next to the tractor and shut the lights off. All is dark and quiet. The clouds hide the stars. In all directions, distant yard lights designate neighboring building sites. Pouring a shot of hot coffee, I contemplate the evening's work.

To the south, where trucks are also awaited, I see the stationary lights of Gunnarson's lifter. Becoming creatures of the night, we've learned to improvise while working in the dark. While others sleep, our nocturnal duties include changing bearings, cleaning mud, finding loose wires and general fixing by flashlight. After starting the tractor, I shine the pickup's lights on the lifter. Searching for a loose chain or worn bearing or anything that might slow us when lifting, I conduct an inspection.

A set of headlights appears east of Kennedy. I watch as they get closer and proceed east, past the turn which would signify our trucks. That one must be Dziengel's. Five minutes later, another is sighted turning off the highway in my direction. I watch carefully. The reflection of the cab's clearance lights identifies the white twin screw. To give the driver my location, I turn the tractor lights on.

Waiting in the idling John Deere, I see his box lights rise. Dumping his dirt, he drives to the water tank to wet down his box. A trick growers have learned is that if the truck boxes are wet prior to loading, the beets will slide out much easier at the pilers. For this reason we park a 500-gallon water tank in the field and pump sever gallons of water into each empty truck box. As he approaches the lifter and swings under the boom, I allow him a couple of seconds downshift before letting out the clutch and heading down the field. Banging loudly, the first beets pound into the bottom of the box. After a small pile is established the noise ceases.

As we move down the field in unison, another set of lights comes down the crossing. The second truck is here. Dumping his dirt and watering his box, he closely follows as the first truck is loaded. Lifti conditions are improving. We can actually load a whole truck without stopping to clean mud. The white twin is loaded. I wave forward, and he drives to the road while the red twin pulls under boom. Looking up, I see the third truck is dumping his dirt.

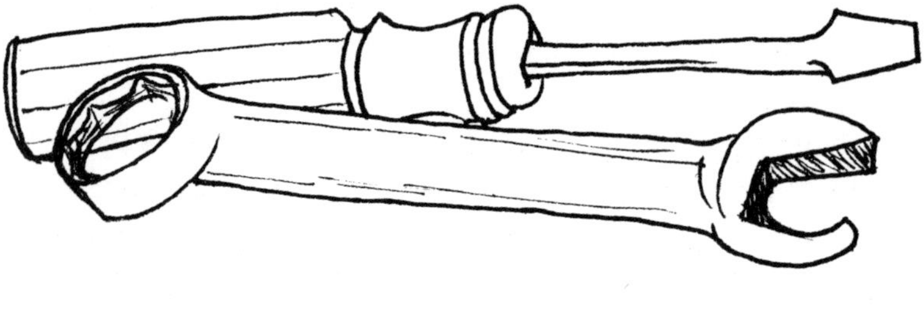

THURSDAY, OCTOBER 9

All was going well until 3:00 a.m., when we started having troubl with the alternator on the tractor pulling the lifter. Frequent stops to jiggle wires and wedge matches in the field connections prolonged its life. Growing dimmer, however, the light finally quit. Removing the alternator, I left the driver in his truck and drove home to the shop and called a local mechanic.

"Hello," he answered sleepily.

"I hate to ask this of you, Alfred," I replied. "But I need a Delco alternator rebuilt. Could you do it for me?"

"Sure, I just charge twice as much this time of night," he laughed.

"That's fine with me."

When I reached the garage he was already waiting. I offered him a cup of coffee but he refused. We spoke in low tones as he worked. Ten minutes ago he was sleeping. Now he was groggily replacing an alternator's brushes. I thanked him and returned to the field.

Anticipating breakdowns, I carry an array of parts to the field, but it seems that the item you don't have is the one that breaks. I should have known better. Next year we will carry a spare alternator.

When I arrived at the field, all three trucks were waiting. We replaced the alternator. The sun was edging over the eastern horizon as the last truck was loaded. We had lost a great deal of time. When you are alloted only twelve hours a day in which to deliver beets, every minute counts. Replacing a bearing here, a stop to clean mud there, or a truck getting stuck, all contribute to losing a few minutes that eventually translate into fewer loads. By shift's end we had lifted only twelve loads. After filling the trucks I arrived home at 2 p.m.

Eight hours later I'm waiting by an open Quonset door. At 10:15 p.m., the drivers arrive separately. The night is cold and black. A good-natured bunch, they smile and greet each other.

"Good evening."

We banter about better working conditions and long hours as each grabs a flashlight and inspects his truck. One truck is a quart low of engine oil. Another driver adds brake fluid. I help pump transmission fluid into the other truck.

"I doubt if we'll lift all night," states one driver. "It's getting pretty cold."

Indeed he is right. The thermometer is already reading twenty-five degrees. If frozen beets are piled they will spoil. I expect Crystal to shut us down sometime during the night.

FRIDAY, OCTOBER 10

Sure enough, we were able to lift only three loads before the drivers were told not to come back. On such a beautiful, calm night it was a shame we couldn't continue. The drivers and I were just getting into the routine of working all night and sleeping during the day. Now we'll be out of sync again. The mercury dipped to twenty degrees. With about nine acres rotobeeted, we'll have to wait and dig them after they thaw and their crowns heal. Fresh beets will have to be topped, leaving these to be dug later.

Getting to bed at 4:00 a.m., I slept until the school bus awakened me at 8:00. Weather will dictate when we can lift again. The telephone pre-recorded message at the Drayton beet plant gives no new announcements. Finally, at 11:00 we're told that American Crystal will start receiving freshly-topped beets at 4:00 p.m. After calling the rotobeeter driver, I go to the beet field. On the way I drive past a neighbor's beets. Mike Gunnarson and Neil Johnson are chiseling and pounding large chunks of frozen beet tops from underneath their rotobeeter. Looking up as I drive by, Mike shakes his head and waves.

After getting the topper going, I return to call the drivers. In the afternoon I lie down but sleep escapes me. The drivers arrive at the usual time: 10:15 p.m.

SATURDAY, OCTOBER 11

We are working on a wetter area of Freda's quarter and the trucks have to be pulled. Perfecting the trick of towing trucks, Robbie insures that not much time is lost. Watching from the Versatile, he can expertly fill a truck without my direction. The night is clear and cold. With steam blowing from our mouths and nostrils, we quickly clean the lifter between loads. If there's a break, Robbie and I retreat to the pickup's warmth. Sipping hot chocolate and listening to Talknet, we wait in the darkness. By 1:30 this afternoon, sixteen loads are delivered to Drayton. It's been a very long day and night. Sleep comes immediately.

The weather is bitterly cold. At 7:00 one of the drivers called. "I'm sorry," he said. "But I won't be able to come to work tonight."

"How come?" I asked.

"Well . . . I haven't been to bed yet and I'm very tired."

"I guess that's your problem, not mine."

"You're right."

"A person needs a little self-discipline and you're going to have to learn to get to bed."

"Well, I know. But I just can't make it."

"Okay, I can't force you to come to work. You're a heck of a good driver and I'd like to have you."

"Well, maybe I can make it."

"Fine. Your truck will be in the yard waiting for you."

SUNDAY, OCTOBER 12

As it turned out the driver never did show and his loaded truck sat in the yard. Tomorrow I'll call a couple of other prospective drivers. I hate to see him quit because of something so foolish as not getting to bed. The clear, cool weather means that it's possible we can combine soon. If so, Robbie can thresh while we dig beets. The temperature continued to drop all night. Expecting each load to be our last, I kept asking the drivers of any word from the scale house.

"I can't figure it out," one replied. "Their thermometer must be stuck at 29. It's 22 in Grand Forks."

As much as I want to lift beets, I don't want them spoiling in the pile either. I hope they know what they're doing. Finally at 5:30 they shut us off. With two empty trucks and one full, we quit. I got home at 6:00 a.m. and went to bed.

This afternoon I took the loaded truck to Drayton and waited an hour-and-a-half to dump at the wet hopper. I'm glad we had only one loaded. The recorded beet report remains the same: no lifting today or tonight until the weather warms.

MONDAY, OCTOBER 13

The no-show driver called this morning.

"If you can still use me, I was wondering if I could have my job back. It won't happen again."

"You're a good driver," I replied. "I'd be happy to have you back. I'll call when we can lift again."

This month's phone bill will show quite a few calls to the same number, the telephone recording at the Drayton plant. This morning's message stated they would receive beets at noon. I made my usual four or five calls to inform the men, and drove to the field to get the rotobeeter going. On the way, I met a neighbor and three of his drivers repairing a lifter stranded on the road. Busy fixing a rear wheel, everyone was in good spirits and we chitchatted about our common problem: beet lifting. It was good to see them maintaining a sense of humor, even dealing with breakdowns.

At 10:00 p.m. the drivers arrived.

TUESDAY, OCTOBER 14

Digging conditions are steadily improving. All three trucks are driving in and out of the field under their own power. Even the tag axle can get loaded and on the road without being pulled. Finding himself without a job, Robbie is reduced to helping me clean the lifter. Finally at 2:30 a.m. I told him to go home and come to work at 1:00 this afternoon. With the sunny weather lately it should go to combine.

The moonless night is cold. After each load I hurriedly clean the lifter and run to the warmth of either the tractor cab or pickup. Listening to all-night talk shows and sipping coffee, I attempt to eat all the lunch my wife has packed. Women have a tendency to take offense if a lunch kit is brought home with anything uneaten. They say such things as, "What's the matter, don't you like my cooking?" "I'll not pack so much next time;" "Maybe you should pack it yourself."

At 5:30 a.m. the night is still pitch black. The lifter has been cleaned and "morning" lunch has just been eaten. Sleepily I sit in the pickup and hear how a somewhat gullible individual from St. Louis got swindled on a get-rich-quick scheme. The talk show host is advising him of his legal rights. On the dash lies a cigarette left by one of the drivers. Having had my eye on it for some time, I light it, take a puff, and inhale deeply, the smoke burning my lungs. I can see why these things are dangerous. With the second puff, I imagine the smoke attacking my mouth, tongue, throat, esophagus, lungs and any other living tissue. Nicotine in the third puff is increasing my heartbeat and blood pressue. I ask myself, "Why am I doing this?" Boredom? Maybe. Trying to manufacture a macho image for myself? Possibly. A victim of advertising? Definitely! Fantasizing that I am a Wyoming cattle drover with a sheepskin coat protecting me from the cold wind, I suck it down to the filter.

Darkness slowly gives way to gray dawn as we continue to lift. The end is in sight. I try to estimate the number of remaining shifts. By 1:30 this afternoon seventeen loads have been hauled to Drayton. Parked in the yard, the full trucks wait for the next shift.

Sometimes trying to get some sleep in the middle of the day can be impossible. Carol and the children have been very cooperative. The kids have tried not to slam doors or shout too loudly. They know Dad's trying to sleep. I keep bribing them by telling them that as soon as beet harvest is over we'll take them to supper in Grand Forks at McDonald's.

"Harry! Get off him!"

I awoke with a jump. Outside I could hear people screaming and kids crying.

"Now what's going on?" I thought. Sleepily lifting the garbage bag from the window, I noticed a family friend, Ivy Bothum, and her four-year-old son, Kevin, had stopped for a visit. For some reason our very docile dog, Harry, had suddenly taken a dislike to Kevin. As soon as Kevin stepped from the car, Harry attacked and pinned the boy on the ground, biting at his head. Kicking the dog off Kevin, Carol ran to the house for me. I dressed and took the dog to a veterinarian at Greenbush and had him put to sleep. His carcass would be checked for rabies.

We felt terrible for Kevin. Ivy took him to the Hallock hospital where he had several stitches sewn on his head and under his eye. We are all fortunate that he wasn't hurt more seriously.

Borrowing a truck from my father-in-law, I send Robbie to start combining. It's good to be getting that done as well.

At 10:15 the cycle of beet lifting begins its repetition.

WEDNESDAY, OCTOBER 15

Orion was already high overhead when we had six loads. It looks like another good shift. All night the drivers were just shadows behind the cab windows. Never stopping to clean mud or talk, we fill the trucks. When they're loaded I wave them to the crossing. The clear, cool night accentuates the stars. The hours tick by surprisingly fast. Gradually a pale, gray eastern sky develops. Yard lights shut themselves off as the grays give way to oranges and yellows. Suddenly, a bright orange crescent bursts on the horizon as Helios whips his steeds into action. At dawn I take inventory of our night's work and estimate the remaining shifts.

En route to their waiting tractors, some of my neighbors who don't raise beets drive slowly past. Preparing for next year's crop, most are applying anhydrous ammonia. I hope next week we'll be doing the same thing. During a break between trucks, I sit in the tractor and feel somewhat gratified by our night's work. Behind me the freshly-disturbed warm earth steams in the cool morning air. Fighting fatigue, I walk around the lifter and breathe deeply.

On the second-to-last load, a clinking sound was heard from the rear end of the red twin screw. Trying to ascertain its source, I walked alongside. Clink . . . clink . . . clink . . . clink . . . My first guess was faulty U-joints, but they seemed sound. Checking the hanger bearings revealed nothing unusual. Again, I walked beside the truck.

Clink . . . clink . . . clink . . . Its perfect rhythm suggested ring and pinion gears. What to do?

"Why don't you take it to Drayton and we'll fill again before taking it to the shop," I said to the driver.

When the shift was over, the loaded truck was taken to the Chevrolet garage.

"What's this going to cost?" I asked the mechanic.

"About two loads of beets. Heh, heh," he answered.

I didn't appreciate his humor.

We dug a very respectable eighteen loads. After replacing quite a few rotobeeter flails and servicing the lifter, I arrived home by late afternoon for three hours' sleep.

Taking a truckload at a time, Robbie continues combining. The wheat is damp but will keep in the wet bin until after beet harvest.

Coming to work this evening, the drivers aren't quite as jovial as they once were. Everyone is getting tired. The red twin is repaired and ready to drive.

THURSDAY, OCTOBER 16

Digging conditions are excellent. We're all glad that the end is in sight. With any luck we can finish this week. Signs of exhaustion are appearing. One driver is swerving all over when I'm trying to load him. Another fell asleep in his truck while waiting at the end of the field. Seeing his clearance lights, I knew he was there. But why didn't he come? I blinked the tractor's lights and called over the CB, but no response. After waiting awhile, I contemplated walking toward him when his lights came on and his truck moved in my direction. Stopping beside the lifter and crawling out, he apologized for dozing off. I couldn't blame him. A twelve-hour shift with our long haul involves a fifteen-hour day.

All night and into the morning we dug. At 9:00 the lifter's

unloading conveyor chain broke. I reacted too slowly and it completely unrolled into the truck box and onto the ground before I clutched. Restringing the fragmented chain required quite a bit of time. Accompanied by much grunting, groaning, lifting, pulling and surprisingly little swearing, the three drivers and I managed to rebuild the heavy chain links and attach them properly to the lifter.

Since many growers are finished, and fewer trucks are being dumped at the plant, I asked the drivers, "Is it okay with you guys if we work a little later into the afternoon, go home, sleep all night and come to work at 6:00 tomorrow morning?"

Three beaming faces looked back at me in agreement. By midafternoon, a couple of bearings went on the lifter and I sent the crew home while I fixed it. Once or twice a summer, I find myself wishing I had never planted a beet. This afternoon was one of those times. After a nineteen-hour day, self-pity and exhaustion rode with me on the way home.

Robbie finished combining wheat this afternoon.

FRIDAY, OCTOBER 17

The full night's sleep felt terrific. Coming at 6:00, the men were in better spirits. Loading the trucks, they headed for Drayton. By midmorning we finished Freda's quarter and moved the twelve miles to dig on Dad's 40. On the way we passed a neighbor's beet field. Half a mile downfield one of their trucks was being loaded. Parked on the headland, their other two sat waiting, the drivers slouched over the steering wheels. I laughed to myself. I had been naive enough to believe that we were the only ones getting tired.

The beets on Dad's are yielding well. The trucks are coming back quickly, perhaps a little too quickly.

"Look what I got!" exclaims one of the drivers.

Waving a traffic ticket out the window, he handed it to me.

"I didn't think the truck could make it to seventy-two miles an hour," I said, reading the citation.

"It's a good thing he caught me empty. I might have been overloaded too."

"Maybe we oughta slow down a bit," I suggested.

Chunks of mud are being thrown from the trucks' tires onto the tar road. While I was cleaning some of the larger ones, Paul Gillie, a loan officer at the local bank, stopped on his way home from work. Wearing a neatly-pressed topcoat and the world's straightest tie, he called out the window, "Get in."

"I'm too dirty."

"Naw, that don't hurt anything."

Dressed in my dirty coveralls, I felt uncomfortable sitting in his immaculately clean pickup. I resisted the temptation to look under the visor for dust.

"I believe the beet growers should set aside a slush fund for cleaning the roads," he said with a twinkle in his eye.

"We're doing the best we can to keep them clean."

We chatted a couple of minutes before parting. It was 4:30 and I was envious that his work day was over. At 7:30 I quit for the day.

SATURDAY, OCTOBER 18

After three-and-a-half weeks, the last day of beet lifting has arrived. The weather is more like September than October. I think we all had our fingers crossed for fear of a breakdown. The trucks are all the same color: dirt brown. The beets are yielding over twenty-one tons, so we are getting more loads than anticipated. We finally finished after dark. Aligning the lifter on the last four rows, I stepped from the tractor and told the truck driver, "This is the round we've been waiting for."

"You bet," he answered. "It'll be good to be done."

The last load went to Drayton after dark, and the men came to the house. Carol served fresh homemade apple pie and coffee as the crew and I sat around the kitchen table. Relieved to be done, we laughed and joked about beet harvest's adventures. The drivers presented their hours and I wrote checks for their wages.

Later that evening I looked out the bedroom window. Sitting near the yard light in their cars, the drivers were celebrating beet harvest's completion with a case of Old Milwaukee.

Having finished combining, Robbie started applying anhydrous ammonia this morning. The last process has begun.

SUNDAY, OCTOBER 19

It was good to be in church today. People who hadn't seen me for awhile asked if I'd been sick.

"Yes," I reply. "Beet fever."

MONDAY, OCTOBER 20

Robbie is applying ammonia on Marion's, north of Kennedy. Containing the highest percentage of nitrogen of any nitrogenous fertilizer, anhydrous ammonia is stored under pressure as a liquid and requires the use of high pressure tanks and metering devices. Under atmospheric pressure, anhydrous ammonia becomes a gas. It is applied by towing a pressurized tank on wheels behind an applicator, with knifelike blades cutting an incision into the soil. The gas is directed through a series of tubes from the tank to each knife and is injected into the soil. Sealing the incision, the soil traps the white ammonia gas. The nitrification process will transform the gas into nitrates, the form of nitrogen used by next year's wheat or sugar beet crop. Taking a forty-foot swath with the applicator, Robbie can cover many acres in a day. I am occupied pulling full anhydrous tanks to the field and drying grain. One of the truck drivers has been hired to do some chisel plowing and discing. It feels terrific having the crop harvested.

TUESDAY, OCTOBER 21

Another beautiful day. We continue to do anhydrous and chisel plowing. The beet ground is getting worked today. To replenish some of the topsoil moisture, people are wishing for an inch of rain. I hope it holds off another week.

WEDNESDAY OCTOBER 22 TO SATURDAY OCTOBER 25

The beet ground is chisel plowed twice, and anhydrous ammonia is applied. Robbie also completes giving all the land to be cropped next year a healthy dose of nitrogen.

MONDAY, OCTOBER 27 TO FRIDAY, OCTOBER 31

It's the last week of work for Robbie. He's put in a good summer and has accumulated many hours. We washed the trucks, the combine and other machinery and put them away for the winter. Radiators were checked for antifreeze and the yard cleaned. It appears as if everything is going to get done before winter sets in.

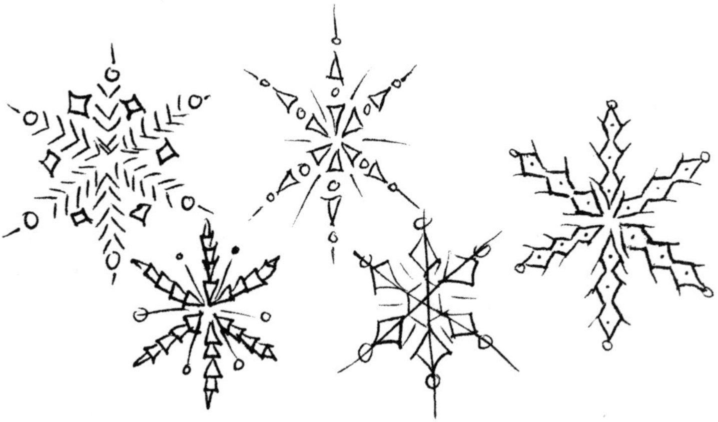

WEDNESDAY, OCTOBER 29

The first snow of the season fell today. It didn't last very long before melting, but you get the feeling that it's only a matter of time before there's an accumulation. The week has been cloudy with a raw northwest wind. The wise rocking chair weather prophets are predicting that when it turns cold it will stay. I believe them.

FRIDAY, OCTOBER 31

The kids got dressed in their Halloween costumes and I drove them to the Kennedy Community Center for a full evening of games. On the way home the radio weather reporters were predicting a large mass of cold air moving in from the Rockies. Accompanied by heavy amounts of snow, it should hit the Red River Valley sometime next week.

SATURDAY, NOVEMBER 8

I'm glad I'm not a deer hunter. Outside a January-like blizzard is raging. Carol's father and brothers are hunting deer in an area known as the Halma swamp in the eastern part of Kittson County. Inside the house I am comfortable and warm. Sipping a cup of hot coffee and watching the snow swirl around the house, I reflect upon the past year. To be sure there are a few regrets. Some decisions I made were incorrect; some proved to be the right thing to do.

Relishing the fact that all the work got done and the anhydrous ammonia was applied this fall, I look forward to the next few months of winter. Becoming re-acquainted with my family after a long beet harvest will be the top priority.

Several good books are awaiting my reading, and this journal will need some extensive fine-tuning before publication. No doubt many basketball games will be played and watched; a trip or two to Fargo's Holiday Inn for a weekend's swimming will be taken.

Farm extension and agricultural informational meetings will be attended. The farm's financial books will be closed for 1986 and an accounting will be given to the bankers. Seed wheat for 1987 will be cleaned; income tax will be paid.

I've been asked a couple of times this week how the year has been. "Good," I reply. But I quickly add, "Any year that you can keep the auctioneer out of the yard is a good year."

The lengthening daylight hours of February and March will once more turn my thoughts toward seeding wheat and planting those delicate beets. May the good Lord give me health to repeat the cycle many more times.

EPILOGUE

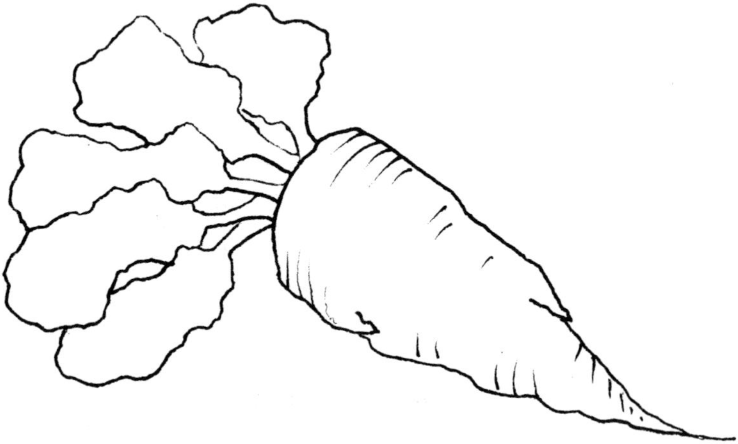

MID-JULY 1988

A year-and-a-half has elapsed since the last entry. As this book goes to press we are locked in the severe drought of 1988 and it seems necessary to include an update on recent events.

The winter of 1986-87 furnished ample snow cover. Warm spring temperatures provided an ideal early-planting season. Spring rains were a little slow in coming but nevertheless gave us an average wheat crop and an outstanding crop of sugar beets. July 14, 1987 was remembered for months, because it was when our last significant precipitation fell. Conditions for grain and beet harvest were the best I've ever experienced. No grain was artificially dried, and because there was no mud to contend with during beet harvest, not one truck was pulled out of the field.

The chronic complainers and perpetual worriers were the first to voice their concern over the dry conditions. Even in the middle of combining, when they should have appreciated the dry weather, they were saying things such as: "It's too dry;" "If it don't rain, there won't be anything next year; "It's too lumpy to plow;" "I can't see how we can put anhydrous on in these dry conditions." Too bad.

Rather than savoring the good working conditions, they were preoccupied with worrying about events weeks and months away.

An easy noon-to-midnight harvesting shift made us ask ourselves constantly if this was really beet lifting. I kept telling myself that things aren't supposed to go this smoothly. I was getting to bed at 3:00 a.m., and actually getting five hours of sleep each night.

A snowless Thanksgiving led to a brown Christmas. Snowmobilers were relegated to traveling groomed trails of grass and dirt in ditches. The unprotected fields were a concern among conservationists. Their concerns proved to be valid.

The spring of 1988 was dry and early with a hot, dry south wind that blew almost daily. With no snowmelt or spring rains, the coulees never flowed. As we were putting the crop in, I optimistically reasoned that it had always rained in the past and it will this year too. In anticipation of that first rain, Robbie and I hurriedly seeded the wheat. Quickly moving from field to field, we took but a few days to seed our 1050 acres of wheat and 40 acres of barley. Sugar beet planting was also completed the last week of April.

Lofting dirt and dust into the air, the relentless wind continued to blow. Cutting visibility and hiding the sun, the air-borne dirt caused motorists to drive with their lights on in the middle of the day.

In an attempt to keep the dirt from blowing across the newly planted beet fields, I "blind cultivated" them. Cultivating young sugar beets is always a tedious task, but doing it before they have emerged is even worse. There is no room for error: a slight shift of the cultivator, left or right, will destroy the newly-sprouted seed. I hoped digging dirt lumps between the rows would provide protection for the young beet plants. In spite of my efforts, the severe winds dislodged soil particles and dashed them along the field's surface. Sandblasting the beet plant's new leaves, the blowing dirt soon left only a thin, decimated stem. Behind each dirt lump or piece of straw residue, small dirt banks were created. Once after a day of gale-force wind, I found the beet cultivator tractor sitting in the field with large banks of dirt extending from its tires. Resembling black snowbanks, they gradually feathered to the northwest, away from the prevailing wind.

Anxiety was etched on people's faces. It appeared in their conversations and surfaced in their body language. One dirt-filled day we took pictures. The sky was cloudless but the sun was nowhere to be seen. I told the kids to carve this spring of 1988 in their memories to tell their children and grandchildren about. I can remember listening in awe to my grandfather telling stories of dirt blowing in 1936. I guess I never doubted him, but wondered how such things were possible. Now I know. The nightly weather forecaster had a new term, "blowing dirt," to describe the next day's weather conditions.

Things were not all gloom and doom, and most people maintained a sense of humor. "If this wind continues," expounded one farmer in the local John Deere dealership, "even all the nonbelievers will be in church." Hmmm . . . Maybe the drought has a purpose after all.

I made daily trips to the beets planted on Hilmer's quarter and the eighty on Section 11. The drought and wind were taking a daily toll on the young, fragile plants. On May 13 another disaster occurred. At 5:45 a.m. I sleepily walked downstairs and looked at the thermometer. 16 degrees! No. It can't be. There must be something wrong with the thermometer. 16 degrees? Can that be possible? It isn't supposed to get THAT cold this time of year, especially during a hot spring. 16 degrees! I was tempted to jump in the pickup and inspect the damage, but I knew what I would find.

"Oh shit," I said. Resetting the alarm for 7:30, I went back to bed, realizing we would have to start over.

When Robbie arrived at 8:00, we hitched the 850 to the multiweeder and headed for Hilmer's to tear up the beets. While Robbie drove the 850, I followed behind in the pickup. On the way I saw a neighbor walking his beet field. I stopped to talk and we crawled on our hands and knees, inspecting his beets.

"Look there," he lamented. Touching a tiny beet plant with a screwdriver, he lifted its limp leaves. "Looks dead to me." He was right. The plant was already losing its green color as the frost crystals melted, removing all turgidity from its leaves.

"I was checking fields at 6:00," he added. "There was frost an inch deep."

"Friday the 13th," I said.

"Yeah," he agreed. He pulled a plant and carefully inspected its black leaves. "Black Friday. With the grain being so poor, I thought I had at least the beets going for us. But now . . ."

We stood and inspected his field, neither one of us speaking. A half-mile south, his four-wheel-drive John Deere approached, pulling a multiweeder. A mile away Robbie had reached Hilmer's with ours. Driving to Hilmer's, I watched as Robbie made the first fifty-foot swath with the multiweeder. A neighbor who doesn't raise beets drove by and almost broke his neck looking at what we were doing. By 1:00 Robbie had torn up all the beets on Hilmer's, and I was making the first rounds replanting. Listening to the radio broadcast from Hallock, I heard the newscaster saying, "It's too early to tell how much damage the frost did this morning."

"Too early to tell!" I exclaimed.

The second planting never takes as long as the first. I upshifted a gear and planted more beets in a day's time than ever before. Four different beet seed varieties had been planted this spring, with little white flags separating them in the field. Ideally the beets would be dug separately in the fall, allowing us to compare each one's performance. Not so on the replant. The same four varieties were planted, but the seed was hurriedly dumped into the planter boxes without any effort to keep them separate. The eighty-acres of beets on Section 11 were also torn up and replanted.

Seed, insecticide, fuel and labor amounted to almost fifty dollars per acre. On 210 acres of sugar beets it's like throwing $11,000 onto the table and rolling the dice again. I don't know why, but I almost enjoyed the replanting. I set my jaw and told myself that I'm going to survive this too. Viewing the spring's adverse conditions as an immense challenge to be undertaken and conquered, I found myself singing a line from a once popular country song, "It might seem crazy, but I wouldn't change a single thing about this life." A few days after replanting, a beautiful half-inch of rain fell, giving the beets a good start.

Later this spring special church services were held and prayers for rain were offered. People commented on how insignificant they felt. "We can do a lot of things toward raising a crop, but we can't make it rain."

At this writing, 850 of our almost-1100 acres of grain have been destroyed. The beets, planted on summer fallow and receiving 2-1/2 inches of rain the last week of May, and another inch a couple of nights ago, are looking good. This fall we will dig twenty-four hours a day. Robbie will take the lifter from noon to midnight and I'll dig from midnight to noon.

Melissa, Alan, Todd and Andrew are growing like weeds. I'm hoping for at least one seven-footer to play for the Celtics. The summer so far has been tremendously busy, both on and off the farm. Bible school, T-ball, basketball camps, Bible camp, swimming lessons, golf and, oh yes, a little farming, make our lives busy and full. Since most of our farm is in summer fallow, the odds for an above-average crop next year are good. Crop insurance will probably cover most of our out-of-pocket expenses. With only a little over two days' combining to do, this fall's pace should be slower; that is, until beet harvest.

ADDENDUM:

Facts and Figures about Sugar Beets

So what happens to Dean's beets after they are brought to the Crystal Sugar Company's hoppers?

The beets are trucked to a sugar-beet processing factory, where they are washed and sliced into what are called "cossettes." The cossettes are conveyed into a diffuser and soaked in water to remove the sugar from the slices. The sugary solution is pumped to a tank, where it is heated and mixed with lime to settle the impurities. The excess lime in the solution is removed by adding carbon dioxide; the impurities are removed by filtering. The resulting liquid, called "thin juice," is put into huge tanks where it is evaporated. This removes the water and causes the sugar to crystallize. The sugar is then packaged and marketed. This beet-sugar processing is carried out in a single operation, resulting in refined sugar.

World-wide, Russia is the top sugar-beet producing country; France is second; the United States ranks third. The Red River Valley in Minnesota and North Dakota is one of the largest sugar-beet growing regions in America. Sugar beets are also grown in California, Idaho, Michigan and Nebraska.

Digging deep into history, it has been unearthed that sugar beets were cultivated by the people of ancient Babylonia, Egypt and Greece. Back in 1744, a German chemist by the name of Andreas Sigismund Marggraf found that sugar derived from sugar beets was the same as that derived from sugar cane. In 1799, a student of Marggraf's, Franz Archard, developed a practical method of removing sugar from the beets. As a result, sugar mills sprang up quickly in Europe and Russia. In 1838 the first successful sugar-beet processing plant was established in Alvarada, California, by E.H. Dyer, an American businessman. Sugar beets didn't become an important source of sugar in this country until after World War I, and their use greatly increased during World War II. Today, sugar beets are second only to sugar cane as a source of sugar.

What about the sugar beet plant itself? It's scientific name is "beta vulgaris," and it's a member of the goosefoot family. It is a biennial plant, meaning it takes two years for the plant to mature. Therefore, the companies that sell sugar beet seeds to growers must leave the plants in the ground for two years. (In areas with cold winters, the beets must be harvested, stored, and replanted in the spring.) The plant then sends out long branches that produce tiny reddish or greenish flowers, from which the seeds are obtained. But the root develops during the first year, so farmers can harvest sugar beets at the end of one growing season.

The plant grows a large, fleshy, tan-colored root; a fleshy stem or crown; and a cluster of dark leaves. The enlarged upper-part of the root is the "beet," and weighs about 1½ to 3 pounds. The root tapers down from the beet. It's long and slender with many little branches. The taproot grows straight down into the earth to a depth of 2 to 7 feet. The crown at the top of the beet is very short, and sends out bunches of brilliant, rich-green, curly leaves, about 14″ long. The fleshy, silvery-white pulp of the sugar beet contains between 13 and 22% sugar, or sucrose. It consists of large water-storing cells and small sugar-containing cells. These cells are arranged in what are called "cambium" rings, which can be seen when the beet is sliced crosswise. They are similar to the annual rings in a tree, as they show growth patterns.

The entire sugar beet plant is useful. The beet, of course, is processed into sugar. The leaves can be cooked and eaten as greens, or incorporated into the field as fertilizer. The leaves and crowns can be fed to cattle, sheep, hogs, and other animals, as can the pulp and molasses that remain after processing.